AUDEL®

Plumbers and Pipe Fitters Library Volume III

Water Supply
Drainage • Calculations

by CHARLES McCONNELL

Macmillan Publishing Company
New York

Collier Macmillan Publishers
London

FOURTH EDITION

Copyright © 1967 and 1977 by Howard W. Sams & Co., Inc.
Copyright © 1983 by The Bobbs-Merrill Co., Inc.
Copyright © 1986, 1989 by Macmillan Publishing Company, a division of Macmillan, Inc.

Macmillan Publishing Company
866 Third Avenue, New York, NY 10022
Collier Macmillan Canada, Inc.

Library of Congress Cataloging-in-Publication Data
McConnell, Charles.
  Plumbers and pipe fitters library.
  Includes indexes.
  Contents: v. 1. Materials, tools, roughing-in—
v. 2. Welding, heating, air conditioning—v. 3. Water
supply, drainage, calculations.
  1. Plumbing—Handbooks, manuals, etc. 2. Pipe-
fitting—Handbooks, manuals, etc. I. Title.
TH6125.M4   1988   696'.1   88-8815

ISBN 0-02-582913-0 (v. 3)
ISBN 0-02-582914-9 (set)

Macmillan books are available at special discounts for bulk purchases for sales promotions, premiums, fund-raising, or educational use. For details, contact:

        Special Sales Director
        Macmillan Publishing Company
        866 Third Avenue
        New York, NY 10022

10 9 8 7 6 5 4 3 2 1

Printed in the United States of America

Foreword

Plumbing and pipe fitting play a major role in the construction of every residential, commercial and industrial building. Of all the building trades, not one is as essential to the health and well-being of the community as the plumbing trade. In addition, the knowledge and craftsmanship of the skilled pipe fitter is required in shipfitting, aircraft, and space vehicles.

Pollution of lakes, rivers, and the aquifer is of vital concern to everyone and is addressed in this three-volume series. Modern technology has created new materials which have revolutionized the piping trades. Materials such as lead and cast iron, used for decades, are being supplanted by plastics.

Plumbing and pipe-fitting installations are governed by codes, regulations, and ordinances established by local, state, and federal agencies. Inspections by licensed inspectors are designed to insure that all rules and regulations governing the work are complied with. Plumbers, and in many areas pipe fitters, are required to pass examinations and be licensed in order to engage in the trade. The health and safety of the population and environmental and ecological concerns demand highly skilled workmen in the building trades. This three-volume series has been written to provide a reference source for those already engaged in the plumbing and pipe-fitting trades and as an aid to those whose intention it is to become a plumber or pipe fitter.

This, the third of three volumes, explains water supply systems, copper tubing, process piping, mathematics, fire protection systems, grooved piping systems, sewage disposal, general plumbing information, questions and answers for plumbers, and concludes with safety on the job.

Contents

Acknowledgments

The author wishes to thank the following companies for their assistance in furnishing information and drawings on their products:

American Society of Automotive Engineers
Asahi/America, Inc.
Hayward Industrial Products, Inc.
Indianapolis Water Co.
Jet, Inc.
Kennedy Valve Co.

Nichols Engineering and Research Corporation
Ridge Tool Co.
Speakman Co.
Suffolk County Water Authority
The Viking Corporation
Victaulic Company of America

CHAPTER I

Water Supply

City water systems obtain their water from several sources—lakes, wells, and reservoirs fed by rivers and creeks. Except in relatively small communities, wells are an uncertain main source of supply. In larger cities, well water is primarily used to assist in meeting peak demand periods and to temper the supply from lakes or reservoirs. For example, the addition of well water at an average temperature of 55° to water from a lake or reservoir that may be 35° or colder warms the colder water and helps eliminate water-main breakages. A typical well and pump used for these purposes is shown in Fig. 1-1.

Cities that depend upon water from creeks or rivers may find that the supply is ample in the spring of the year, due to melting snow and spring rains, but not during summer and fall. Water from these sources must be conserved and stored for future use. Reservoirs are built for this, and a dam is built at the lower end of the reservoir to impound the water. Reservoirs are designed to maintain a certain maximum level; flood gates are built into the dam to permit the release of excess water if the level becomes too high. Reservoirs also assist in flood control by storing water that would otherwise cause flooding for slow release later. Besides storing water for future use, they may also serve as recreational spots. A typical reservoir of this type is shown in Fig. 1-2.

Fig. 1-1. A well and pump for a city water system. *(Courtesy Indianapolis Water Co.)*

Water stored in the reservoirs is "raw" water, unfit in almost all cases for human consumption. The water utility, starting with the raw water, filters and chemically treats the water to make it potable. Fig. 1-3 shows the various stages of treatment and testing as raw water from a reservoir is processed and delivered to its consumers.

Raw water entering a treatment plant usually contains minerals,

Fig. 1-2. Aerial view of Morse Reservoir, one of several reservoirs serving the Indianapolis area. *(Courtesy Indianapolis Water Co.)*

silt, sand, and other biologic organisms. In normal processing, little is done to remove the dissolved salts such as limestone; the emphasis is on removing the suspended matter. As the water enters the plant a coagulant, aluminum sulphate (alum), is added; this chemical combines with water to form aluminum hydroxide, a visible-sized gelatinous mass called "floc." Immediately after the alum is added to the incoming water, the mixture enters a large mixing basin. Large paddles rotate slowly in the water to thoroughly mix the chemical with the water. After mixing, the water moves into the settling portion of the basin. There is no agitation in this portion of the basin; the water is moving very slowly, and the floc particles, with the added weight of the suspended material in the water, settle to the bottom. This process removes 95 percent of the foreign particulate matter. At the end of the basin, weirs (dams) at the water's surface allow the upper-level, higher-quality water to pass on to the next stage of treatment, the filtering basin.

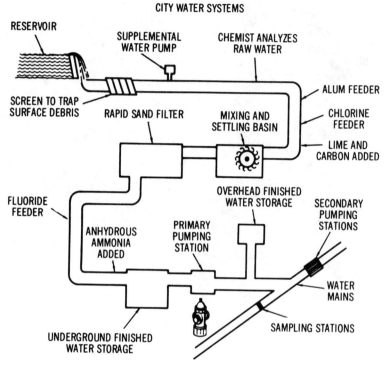

Fig. 1-3.   Steps in producing and distributing safe water.

A filter is another concrete box with a series of pipes that brings the settled water to the filter. Within the filter lies an underdrain collection system for the filtered water. Above the collection system lies 24 to 30 in. of fine sand and anthracite (fine coal) supported by five layers of graded gravel. Water from the settling basin passes down through the filter sand and gravel. Here all but the finest colloidal particles are removed, and clear water leaves the filter bed. A certain percentage of harmful bacterial is also removed in the filtering bed. A typical filter bed is shown in Fig. 1-4.

After filtration and clarification, the raw water enters the chemical treatment tank for sterilization with chlorine. Chlorine is added to the raw water before it enters the mixing basin. The chlorine is thoroughly dispersed through the raw water in the mixing basin, and sufficient contact time is provided in the mixing and settling

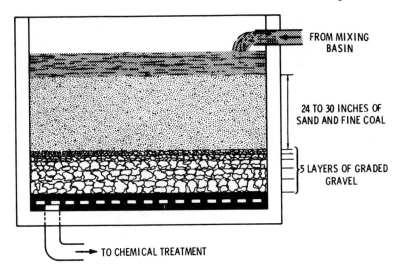

FROM MIXING BASIN

24 TO 30 INCHES OF SAND AND FINE COAL

5 LAYERS OF GRADED GRAVEL

TO CHEMICAL TREATMENT

Fig. 1-4.   Details and construction of a sand filter bed.

basins and filters to insure an effective bacteriological kill. The chlorine oxidizes certain inorganic and organic matter; it also acts as a decolorizer and removes some of the objectionable tastes and odors. The amount of chlorine used is in direct relation to the pollution level of the raw water. The maximum residual level of chlorine in the potable water leaving the plant is 1 part per million. Anhydrous ammonia is another chemical regularly used in water treatment. It is added as the water leaves the sand filter, and it combines with the chlorine to reduce taste and odor problems.

In some water systems another chemical, fluoride, at a level of 1 part per million is added to reduce the incidence of tooth decay in children. A fluoride feeder is shown in Fig. 1-5. The water that entered the plant as raw water has now been screened, filtered, and chemically treated to render it safe for human consumption.

Water Storage

Water that has been purified must be stored for use during periods of peak demand. Large underground finished water reservoirs similar to the one shown in Fig. 1-6 are often used for this purpose.

Fig. 1-5.   A chemical feeder controls the addition of fluoride to water.

In addition, water towers similar to the one shown in Fig. 1-7 are used, not only to maintain a reserve supply for use during peak demand periods but also to maintain pressure in the distribution mains. Water from the distribution mains can be pumped into the tower at night, during periods of low demand, for use during the

Fig. 1-6. An underground finished water storage reservoir. *(Courtesy Indianapolis Water Co.)*

day at peak demand. The stored water helps maintain a constant even pressure. A column of water one foot high exerts a pressure of .433 lbs. per sq. in. at its base. Therefore, the water in a tower 100 ft. high will exert 43.3 lbs. of pressure through the vertical riser to a point at ground level. If the distribution main is 5 ft. below ground level, the tower water will exert 45.465 lbs. of pressure on the main. Water towers in larger cities are usually placed at strategic points to insure a plentiful supply during periods of peak demand.

Distribution Mains

When the raw water has been "finished," or purified, it must be sent to the users through underground distribution mains. Over a period of years, various materials have been used in the manufacture of underground water mains; cast iron has proven to be the most durable. Cast iron rusts on the surface but protects itself against the

Fig. 1-7. **An overhead water storage tank.** *(Courtesy Indianapolis Water Co.)*

continued rusting that would, in time, destroy the metal. Indeed, when cast iron rusts on the surface, the granular rust coating adheres strongly to the metal and protects unrusted parts from further corrosive action. A type of cast iron called ductile iron is widely used for water mains; it has the advantage of being more flexible, less

Fig. 1-8. **Workmen installing water main pipe.** *(Courtesy Suffolk County Water Authority)*

brittle, than ordinary cast-iron piping. Fig. 1-8 shows workmen installing cast-iron water mains.

In order to maintain a constant pressure on distribution mains throughout the area served by the water utility, primary pumping stations, similar to the one shown in Fig. 1-9, are used to start the water on its way through the mains. Secondary pumping stations are located at strategic points to boost the pressure when necessary to insure a steady flow throughout the system. Fire hydrants are placed at regular intervals on the distribution mains, making water under pressure readily available for use in emergencies.

Water Testing

A modern water utility makes many tests daily, both on raw and finished water. Finished water is tested at widely scattered points

Fig. 1-9.   A primary water pumping station. *(Courtesy Indianapolis Water Co.)*

throughout the system to maintain the highest standards of purity. Fig. 1-10 shows a chemist examining a water sample.

Chemicals used regularly but not necessarily continuously in water treatment are:

Copper sulphate to control algae and other biologic organisms.
Lime to increase the alkalinity and pH* during periods of high stream flow.
Carbon for taste and odor control.
Sulphur dioxide as a dechlorinating agent.

Wells

The use of surface water, other than that collected on roofs or other controlled areas and stored in cisterns for rural domestic use, presents many difficulties and should be avoided. Wells, of course, may be used. In terms of construction, they may be classified in the following groups:

*pH—a measure of the acidity or alkalinity of water.

1. Dug.
2. Bored.
3. Driven.
4. Drilled.

Fig. 1-10. A chemist tests water samples. *(Courtesy Indianapolis Water Co.)*

Dug Wells

A dug well is constructed by excavating a shaft, generally manually, and installing a casing where needed. Dug wells are used extensively for domestic water supplies. They are generally not very deep because they cannot readily be sunk far enough below the water table. Most are less than 50 ft. deep. They generally yield only small supplies of water from water-bearing materials of rather low permeability near the top of the zone of saturation.

Dug wells are necessarily relatively large in cross-section, and they have correspondingly large storage capacity for each foot that they extend below the water table. Because of their shallow penetration into the zone of saturation, many dug wells fail in times of drought when the water table is reduced. A dug well is illustrated in Fig. 1-11.

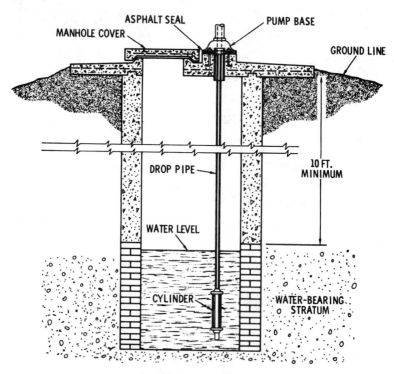

Fig. 1-11. Construction details of a dug well.

Bored Wells

A bored well is constructed by boring a hole with a hand or power auger and installing tile or other casing. Bored wells range in diameter from only a few inches when hand-operated augers are used, to 4 ft. or more when power augers are used. Like dug wells, bored ones do not extend into or through hard rock, and most of them are not sunk far into the zone of saturation. To a great extent bored wells resemble dug wells in that they generally have relatively small yield and are quite often affected by drought.

Driven Wells

A driven well is constructed by driving a pipe (usually equipped with a well point and screen) with a maul, drive donkey, or pile driver. Driven wells are confined to localities where water-bearing sand or fine gravel lies at comparatively shallow depths, and where there are no intervening hard rocks or boulders that would prevent driving the pipe. Under these conditions, driven wells can be constructed rapidly and at small cost. The pipes used are normally 2 in. or less in diameter. These wells are usually pumped by suction from pumps located at the top of the pipe. Wells of this type are likely to be impractical if the sand or gravel has low permeability, but if the permeability is high, the well may be plentiful.

Drilled Wells

A drilled well is constructed by making a hole with a drilling machine and installing casing and a screen where needed. The excavating may be done by percussion or rotary tools or by jetting, and the materials may be brought up by means of a boiler or hollow-drill tool or by a hydraulic process.

In the percussion well-drilling method, a heavy drill bit is suspended in a prepared hole in the ground by means of a wire rope attached at the upper end to the walking beam of a derrick. The bit is alternately lifted and dropped by the machinery, pounding the earth and rock into small fragments for subsequent removal from the hole.

Drilling methods have a great advantage over digging, boring, and driving methods because they are adapted for sinking the holes

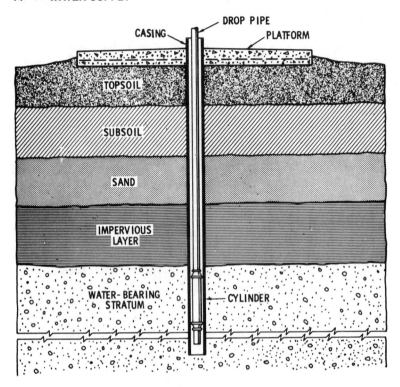

Fig. 1-12.   Cross-sectional view of a drilled well, showing its path through several layers of soil.

to water-bearing beds that may be far below the water table. Drilled wells are more apt to tap water supplies that might not be recoverable with wells of other types, and as a rule they have larger yields and are less affected by drought. A drilled well is shown in Fig. 1-12.

City Water Sources

The sources from which city water is obtained for domestic purposes are numerous, such as wells, springs, lakes, rivers, and rainfall. There are several methods of getting the water to the point of supply, such as:

1. Street pressure.
2. Pneumatic.
3. Electric.

Street Pressure System

The source of supply of the street pressure system is the street main into which water is furnished under pressure by the water company. A *service pipe* is run from the main to the dwelling and connects to the *supply pipe*, and from there suitable branches are run to the various fixtures, as shown in Fig. 1-13. Although this system is mechanically simple, it sometimes has disadvantages that depend

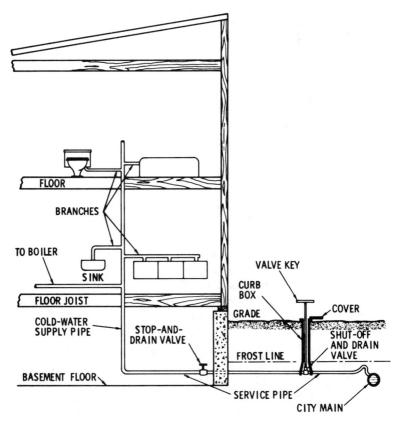

Fig. 1-13. A typical cold water supply line.

on the pressure. Where the pressure is excessive, the piping and fixtures are subjected to needless pressure, causing excessive load on the valves and tending to increase leakage; if the pressure is too low or variable, the water may not flow to the outlets on the upper floors.

Sometimes, when the pressure is excessive, a reducing valve is placed on the supply pipe to protect the line and fittings from unnecessary pressure. It should be understood that, in any system, the installation must withstand not only the working pressure but also momentary shocks due to water hammer caused by the quick closing of faucets. The weight of the pipe used must be adequate for the working pressure. Because of the enormous strain possible from water hammer, and because of general deterioration from use, it is advisable to use pipe that, when new, will have a pressure rating that is far in excess of working pressure.

Pneumatic System

The word *pneumatic* is defined as pertaining to devices that make use of compressed air. Accordingly, the pneumatic system of a water supply makes use of compressed air to elevate the water to the various outlets in a building. The apparatus required consists essentially of a closed cylindrical steel tank and a pump for raising the water from the source of supply and forcing it into the tank. When there is no water in the tank, it contains air at atmospheric pressure, or 14.7 lbs. per sq. in. In operation, the pumping of water into the bottom of the tank will compress the air after the opening into the tank has been closed. As the air is lighter than water, it is compressed into the space above the water. As the water level rises at each stroke of the pump, the air becomes compressed more and more in the top of the tank until it finally reaches the desired point of compression. The very strong pressure pushes the water out from the tank and through the pipes to any part of the house or grounds, where it is then ready to flow. Air is very elastic and acts much like a wound-up spring. Its force becomes less when the volume of water decreases and the air space expands. By increasing or decreasing the amount of air put into the tank, and also the pressure, pneumatic systems will meet the requirements of various locations requiring either a high or low pressure.

There should be an air-charging device provided so that ad-

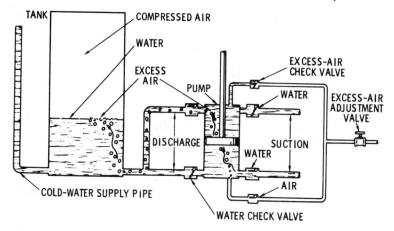

Fig. 1-14. A pneumatic water system.

ditional air is pumped into the tank at each stroke of the pump, if desired. This device usually consists of an air valve and a check valve attached to the cylinder head, as shown in Fig. 1-14. A small quantity of air is forced into the tank at each stroke of the pump to maintain the tank pressure. The amount of air pumped may be controlled by adjusting the air valve. Normally, and for ordinary elevations, little or no air is required.

Power Pumps

In selecting a water pump, consideration first must be given to its duty and to whether the pump possesses sufficient capacity to supply the home with the desired quantity up to the yield of the well. Pumps employed for water pumping may be classified with respect to their cycle of operation as:

1. Plunger or reciprocating.
2. Turbine.
3. Centrifugal.
4. Rotary.
5. Ejector.

They may also be divided into *shallow-* and *deep-well* pumps. Shallow-well pumps are sometimes referred to as *lift* or *suction* pumps. The plunger-type pump is occasionally termed a *positive-displacement* pump. The rotary pump is also a positive-displacement pump. Such pumps will continue to build up pressure as long as they continue to operate.

Centrifugal and ejector pumps develop pressure by centrifugal force and cease to build up pressure beyond a given limit. This limit depends on the design and speed. The pressure beyond which these pumps will not force water is termed the *shutoff* head. The discharge from centrifugal pumps can be regulated by means of a valve in the discharge pipe, but it should not be completely stopped in this manner.

Other mechanical means of lifting water are air-lift and air-displacement pumps, chain pumps, propeller and screw-type pumps, hydraulic rams, and siphon pumps. Air-lift and air-displacement pumps are usually not efficient and are therefore not often used. Chain pumps are not considered as sanitary as other force pumps and are also inefficient. The propeller or screw-type pumps are used for lower lifts and larger volume than are ordinarily required for domestic use. Hydraulic rams obtain their power from the water supply, and are wasteful, although completely automatic. The hydraulic ram is probably the most inefficient of the mechanical means mentioned above, if the quantity of water delivered is compared with the quantity of water required to supply it.

Turbine Pumps

The turbine pump is very simple in construction and consists essentially of one or more circular discs, as shown in Fig. 1-15. The outer edges of the discs are fitted with a series of equally spaced vanes or blades. The discs are firmly mounted on a drive shaft, while the assembly is surrounded by a close-fitting envelope or case. In operation, the disc rotates, with very little clearance, in a channel formed by the casing liners. Rotation of the disc forces the water to move with it, carrying the water from the inlet to the discharge line. Turbine pumps may be mounted with the shaft either vertically or horizontally.

The horizontal-type pump is often installed in shallow wells, but in deep wells the motor is mounted vertically and the shaft extends down below the water level. In some wells, one impeller

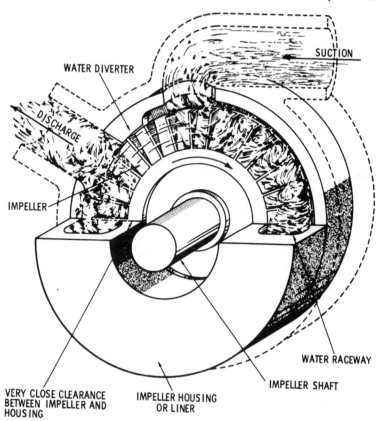

WATER DIVERTER

SUCTION

DISCHARGE

IMPELLER

VERY CLOSE CLEARANCE
BETWEEN IMPELLER AND
HOUSING

IMPELLER HOUSING
OR LINER

IMPELLER SHAFT

WATER RACEWAY

Fig. 1-15.   The working principles of a turbine water pump.

may not develop sufficient pressure to deliver water to the surface
of the well and against the pressure of the tank. In such cases,
additional impellers are used in sufficient numbers to develop the
desired pressure. Turbine units equipped with two impellers are
termed *double-stage* turbine pumps, and if there are many impellers, they are called *multi-stage* turbine pumps.

Centrifugal Pumps

A centrifugal pump is different from a turbine pump. Instead of
having the open blades mounted on the outside rim of a disc, in a
centrifugal pump a number of vanes radiate from the hub, as in Fig.

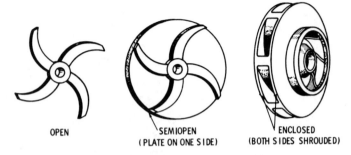

OPEN   SEMIOPEN   ENCLOSED
(PLATE ON ONE SIDE)   (BOTH SIDES SHROUDED)

Fig. 1-16. Various forms of vanes used in centrifugal pumps.

1-16. The impeller operates in a close-fitted housing adapted to the particular form or forms of vanes used. The water enters the impeller from one or both sides at the hub, and is thrown out by centrifugal force. The casing has a volute (snail-like) passage extending around the impellers. This passage begins very small and increases in cross-sectional area to the discharge. Water moving out through the impeller creates a vacuum at the center.

To use centrifugal force in a pump, there must be an impeller, a pump body, and a source of water. When the motor is turned on and the impeller starts to rotate at high speed, water is forced from the center of the impeller, out through the edge of the ports through an area known as the volute passages, and into the pump body. As this water is thrown out, atmospheric pressure on the surface of the well forces water up through the suction pipe into this area to relieve the partial vacuum.

Centrifugal pumps of the conventional type must be primed. They are not designed for high-suction lifts and are therefore usually set near the water's surface and operated with foot valves that keep water in the suction pipe. The pressure at the discharge end depends on the speed and diameter of the impeller.

The volume depends to a great extent on the width of the impeller and the size of the water passage. Centrifugal pumps are water-lubricated and are thus subject to excessive wear if the water carries abrasive material. If the water is free from such material, centrifugal pumps will last a long time. Centrifugal pumps, like rotary pumps, may be set with the shaft either horizontal or vertical. They also take their designations from the position of the shaft.

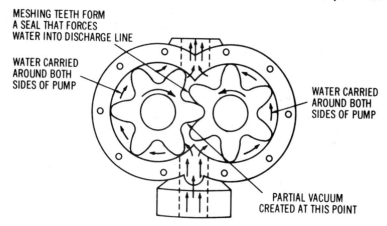

MESHING TEETH FORM
A SEAL THAT FORCES
WATER INTO DISCHARGE LINE

WATER CARRIED
AROUND BOTH
SIDES OF PUMP

WATER CARRIED
AROUND BOTH
SIDES OF PUMP

PARTIAL VACUUM
CREATED AT THIS POINT

Fig. 1-17.   The mechanism of a typical rotary pump.

Rotary Pumps

The rotary pump is frequently called a *gear* pump because of its design, which is in the form of two gears. These gears mesh together inside the pump housing, as shown in Fig. 1-17. This type of pump is seldom used in domestic water-supply systems, but is used a great deal in pumping oil or other lubricating liquids that are free from abrasive materials. The liquid enters at the bottom (Fig. 1-17), fills the space between the teeth, and is carried out around the top, where it is forced out when the teeth mesh with those of the opposite gear. The pump delivers a steady flow of liquid without pulsation, and since it is a positive-action type, it will operate against any pressure the equipment allows.

Ejector Pumps

The ejector, or jet, pump is relatively simple in construction and has a high capacity under low-pressure conditions. It is used in both shallow and deep wells to a depth of 80 ft. or more. The pumping mechanism usually consists of a vertical centrifugal or turbine pump, of the type previously discussed, operated in connection with a jet which is set up in the well.

The ejector pump is actually two pumps working together, one discharging into the other. This is shown in Fig. 1-18. A cross-

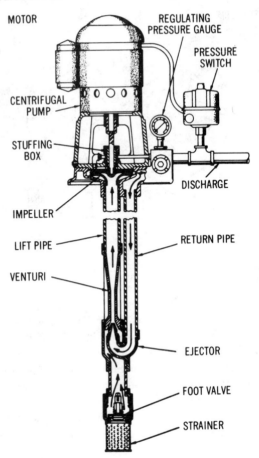

MOTOR

REGULATING
PRESSURE GAUGE

PRESSURE
SWITCH

CENTRIFUGAL
PUMP

STUFFING
BOX

DISCHARGE

IMPELLER

LIFT PIPE

RETURN PIPE

VENTURI

EJECTOR

FOOT VALVE

STRAINER

Fig. 1-18.   The principle of operation of an ejector or jet-type pump.

sectional view of the pump shows that, when properly primed and running, it delivers a given quantity of water to the surface by suction. At the discharge end, some water is forced into the pressure-tank line and some water returns to the jet, or ejector, nozzle.

Submersible Pumps

In a water system of this type, the entire pumping unit is submerged in the water in the well. This relatively new addition to the pump family is designed for deep-well installations and consists essentially of a pump and motor built together into a long slender unit. The motor is placed directly below the pumping unit, and a waterproof electric cable furnishes power to the motor.

It should be kept in mind that a submersible pump requires an ample supply of water and that the well must be free from sand in suspension. As with other centrifugal pumping systems, a quantity of sand will quickly ruin both the pump and the motor. The submersible pump is more sensitive to sand and grit, mainly because of the great precision necessary in its construction.

CHAPTER 2

Working with Copper Tubing

There are four types of copper tubing in common use.

Type K: soft (rolls) hard (lengths)
Type L: soft (rolls) hard (lengths)
Type M: hard (lengths)
DWV grade: hard (lengths)

As can be seen in Table 2-1 the difference between these types of tubing is in the wall thickness of the tubing.

Table 2-1.   Types and Uses of Copper Tubing

Copper Water Tube					
Nominal Size (IPS)	Actual O.D. Size	Type K Wall Thickness	Type L Wall Thickness	Type M Wall Thickness	DWV Wall Thickness
¼″	.375	.035	.030	.025	
⅜″	.500	.049	.035	.025	
½″	.625	.049	.040	.028	
⅝″	.750	.049	.042	.030	
¾″	.875	.065	.045	.032	
1″	1.125	.065	.050	.035	
1¼″	1.375	.065	.055	.042	.040
1½″	1.625	.072	.060	.049	.042
2″	2.125	.083	.070	.058	.042

Standard Colors of Copper Water Tube

Type K—green; Type L—blue; Type M—red; DWV—yellow. Type of tubing is marked on tubing in above colors. Type ACR (air-conditioning and refrigeration) tubing is equivalent in size to Type L and is sold with sealed ends to prevent moisture entering tubing.

Recommended Usage

Type K—Underground and interior water service.
Type L—Interior water service, panel heating, underground snow melting systems, irrigation and sprinkling.
Type M—Interior heating and non-pressure applications.
Type DWV—Drainage, waste, and vent piping. *Above-ground use only.*
Type ACR—Air conditioning and refrigeration.

Types K and L are sold in both hard temper (lengths) and soft temper (rolls). Types M and DWV are sold only in hard temper (lengths).

A copper-cutting tool is shown in Fig. 2-1. The de-burring tool shown in Fig. 2-2 can be used to remove the burrs formed in cutting. End (A) will remove the inside burr; end (B) will remove the outside burr.

There are three ways by which copper tubing and fittings can

Fig. 2-1.   A cutter used to cut copper tubing. *(Courtesy Ridge Tool Co.)*

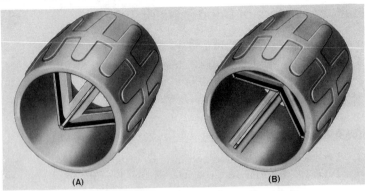

(A)   (B)

Fig. 2-2.   A de-burring tool. *(Courtesy Ridge Tool Co.)*

be joined. These joints can be made using flare type fittings, ferrule type fittings, or soldered or "sweat" joints. Flared joints are used primarily for water service connections between a building and the water service main, using $\frac{3}{4}$-in., 1-in., or $1\frac{1}{4}$-in. soft copper tubing.

Ferrule type fittings can be used with either soft or hard copper tubing, are primarily used with $\frac{3}{4}$ in. and smaller tubing, and should only be used in accessible locations.

Soldered or "sweat" joints require skilled workmanship and knowledge of the correct type of solder to use for a particular application.

There are two basic types of solders used by the piping trades:
1–Soft solders containing various percentages of tin and antimony and tin and lead.
2–Hard solders, alloys containing various percentages of silver, copper, and phosphorus, commonly termed silver solder.

Copper tubing has many uses, ranging from water service piping from water mains or wells to points of usage in homes, factories, processing industries, and service industries. It has been common practice in past years to use solders and solder/flux combinations containing lead and tin for making joints in copper tubing and fittings in many of these applications. Extensive studies have determined that lead in solders used for making joints in copper tubing and fittings can be dissolved or leached into water passing through the tubing and thus poses the threat of lead poisoning to persons drinking or ingesting this water. The U.S. Congress in an amendment to the Safe Drinking Water Act, Public Law 99-339, effective June,

1988, prohibited the use of solders and fluxes having a lead content in excess of 0.2 percent ($\frac{2}{10}$ of 1 percent) in the making of joints or fittings in any public or private potable water system.

95/5 solder—95 percent tin and 5 percent antimony—meets the requirements of a lead-free solder and can be used to make joints in potable water supplies.

Sweating Copper Tubing and Fittings

There are a few simple rules to follow when sweating joints in copper tubing. Following these rules will result in perfect leak-free joints:

1. The male end of the tubing and the female end of the fitting must be clean and bright.
2. The piping being joined *must* be dry. It is impossible to sweat a joint properly if there is water in the joint.
3. Use plain 50/50 or 95/5 solder, whichever is best for the particular job, and a noncorrosive lead-free soldering paste flux. Do not use acid core or rosin core solder.
4. Heat must be applied to the right place on the fitting. Capillary action will then pull the solder into the joint.
5. When working on a closed system of piping (i.e., one in which the piping being connected is closed by a valve or other fitting at the ends), a valve or fitting must be opened. When heat is applied to a closed system of piping, pressure is built up. If the built-up pressure is not relieved, it will cause the melted solder to blow out of the newly made joint and cause a leak.

Clean the male ends of the copper tubing and the female ends of the fitting with sand cloth or fitting brush. Coat the cleaned ends with solder paste. Join the pieces together and apply heat from the torch. Fig. 2-3 shows where the heat should be applied. Apply the end of the solder to the joint; when the joint is at the right temperature for sweating, the solder will flow into the joint. Play the heat from the torch all around the fitting as shown in Fig. 2-3. Follow the heat around with the solder, making sure that the solder has flowed into the joint all the way around. Remove the heat and allow the joint to cool.

The use of lead/tin solders and fluxes containing lead/tin may be desirable when making joints in piping systems carrying other

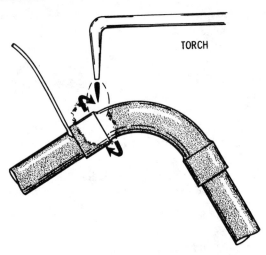

TORCH

Fig. 2-3. Sweating (soldering) copper pipe and fittings. Heat should be applied on the back side of the fitting to draw the solder into the joint.

than potable water. The risk of lead being leached into the material carried by the piping system remains and should be considered.

Silver Brazing Copper Piping and Fittings

The silver brazing process, also called "silver soldering," for joining copper tubing and fittings, is very different from the sweating (soft soldering) method. Silver brazing rods are alloys usually composed of copper, silver, and phosphorus with a silver content of approximately 15 percent. Silver brazing rod will start to melt at 1185°F and become completely liquid and free flowing at 1300°F.

A torch capable of heating copper tubing and fittings to a heat sufficient to melt the silver alloy brazing rod must be used. The copper pipe and fittings must be heated to a dull cherry-red color in order for the silver alloy to melt and be drawn into the joint by capillary attraction.

In most cases when making copper-to-copper silver soldered joints a flux will not be needed and should not be used if the melted alloy flows freely around the joint.

When joining copper tubing to brass fittings and valves, using silver alloy brazing rod, a flux will be needed and only fluxes made for silver brazing should be used. Welding type glasses or goggles, Shade 4 or 5, should be worn while making silver soldered joints.

When the joint, both pipe and fitting, is heated to the cherry red color, the silver alloy rod when applied to the joint will melt and as the torch flame is played around the circumference of the joint the alloy will follow. When the joint is filled the flame should be removed and the joint allowed to cool or "set up" before moving it.

More detailed information on silver brazing will be found in Chapter 1 of Volume II of PLUMBERS AND PIPE FITTERS LIBRARY.

CHAPTER 3

Process Piping

The installation of process piping provides a large part of the work performed by plumbers and pipe fitters. Manufacturing processes used by pharmaceutical and chemical industries, electro-plating works, computer-chip companies, and many other industries involve the use of distilled water, de-ionized water, acids, alkalies, gases, as well as other chemical compounds and substances. Many of these materials are highly corrosive and attack the pipe, fittings, and valves through which they pass. The atmosphere in the area of use and storage is often highly corrosive also, resulting in the piping and the valves being attacked from the outside as well as from the inside.

Until the advent of thermoplastics, the only materials available for process piping were steel, nickel (stainless steel), lead, and glass. Corrosion-resistant thermoplastic systems are rapidly replacing metal process piping systems. Thermoplastic process piping systems offer not only corrosion-resistant piping but freedom from contamination.

The most widely used thermoplastics are:

PVC—(polyvinyl chloride).
CPVC—(chlorinated polyvinyl chloride).
PP—(polypropylene).
PVDF—(polyvinylidene fluoride).

PVC

Polyvinyl chloride is inert to attack by most strong acids, alkalies, salt solutions, alcohols, and many other chemicals. PVC imparts no tastes or odors to material handled. It cannot react with materials carried, nor act as a catalyst. All possibility of contamination or chemical process changes are eliminated. PVC Schedule 40 pipe provides high tensile and high impact strength and it is lightweight, 40 lbs. per 20-ft. length as against 250 lbs. per 20-ft. length of 4-in. steel pipe.

Industrial fumes, humidity, salt water, weather, or underground conditions will not damage PVC pipe or fittings and it is immune to galvanic or electrolytic attack. The mirror-smooth, nonwetting interior surface of PVC pipe assures low head and friction loss both initially and after prolonged service. Tight and dependable joints in PVC pipe and fittings can be made using solvent welding techniques. A PVC True Union Ball Valve is shown in Fig. 3-1. The

Fig. 3-1. A disassembled all-plastic True Union Ball Valve. *(Courtesy Hayward Industrial Products, Inc.)*

Fig. 3-2. A flanged three-way valve. *(Courtesy Hayward Industrial Products, Inc.)*

collars at the ends of this type of valve can be loosened to permit removal of the working parts for cleaning or repair if needed. A flanged three-way PVC valve is shown in Fig. 3-2. A PVC valve designed for pneumatic or electric actuators is shown in Fig. 3-3.

CPVC

Chlorinated polyvinyl chloride piping has two highly desirable characteristics:

1. Outstanding mechanical strength at high temperatures.
2. Relatively low cost.

Fig. 3-3.   Ball valve designed for either pneumatic or electric actuators. *(Courtesy Hayward Industrial Products, Inc.)*

For pressure piping applications it is recommended for temperatures as high as 180°F; and Type 4 Grade 1 CPVC has a working temperature range of 200°F and is recommended where hot corrosive liquids are being handled. The National Sanitation Foundation has approved CPVC pipe for use with potable water. Some typical applications of CPVC pipe: chemical production, waste treatment, pulp and paper manufacturing, metal plating, fertilizers, ore refining, detergents, and semi-conductors.

CPVC pipe and fittings can be joined by solvent cementing or hot-gas welding. Some typical CPVC products are: ball valves, ball check valves, unions, bushings, faucet parts, hangers, caps, plugs, needle valves, and reducers. CPVC is also essentially immune to solvation or direct chemical attack by virtually all inorganic reagents, aliphatic hydrocarbons, and alcohols.

CPVC is capable of handling mixtures of corrosive solutions of varying concentrations. It is inert and will not initiate catalytic activity in the contained fluid. The low thermal conductivity of CPVC reduces moisture condensation on water lines.

PP

Polypropylene Type 1, a member of the polyolefin family, is one of the lightest plastics known. It is highly crystalline, thus is strong and hard. It possesses excellent chemical resistance to many acids, alkalies, and organic solvents and has excellent dielectric qualities. It is reinforced with glass, an anti-oxidizing agent, and UV (ultra-violet) stabilizer, adding overall strengthening to the polypropylene and extending its upper temperature limits. PP Type 1 can be joined using either the socket fusion method or the butt fushion method.

PVDF

Polyvinylidene Fluoride Type 1 is a high molecular weight fluoro-carbon that has superior abrasion resistance, chemical resistance, dielectric properties, and mechanical strength. PVDF maintains these characteristics over a temperature range of $-40°F$ to $250°F$. PVDF is highly resistant to wet or dry chlorine, bromine and other halogens, most strong acids and bases, aliphatics, aromatics, alcohols, and chlorinated solvents.

Thermal Welding

Joints in piping and/or fittings in both PVC and CPVC can be made by solvent welding but polypropylene's crystalline structure resists even the strongest of organic solvents. Since the pipe is impervious to the swelling effects of surface primers and it resists stress crack corrosion caused by polar solvents, heat fusion techniques must be used to join polypropylene systems. Butt fusion is the easiest, most cost-effective, and dependable way of joining polypropylene system components. When butt fusion welds are used, the user can visually inspect the integrity of the joint, whereas socket fushion welds are internal and cannot be inspected visually. In addition, tests indicate that welds created by the butt fusion method are stronger than the pipe itself.

Butt Fusion Welding

The step-by-step process of butt fusion welding is shown in Fig. 3-4. The pipe is cut perfectly square (a), and secured into the Proweld tool. The planing unit shaves off both ends of the pipe (b). The planing unit is then removed and the heater plate, Fig. 3-5, which has been pre-warmed, is lowered into place (c). Full welding pressure is applied (d) to melt the skin of the faced area. The pressure is then reduced to the recommended melt pressure, applying only minimum pressure, enough to assure full surface contact. When a symmetrical bead appears (e), separate the pipes with a quick, smooth motion, remove the heating mirror, and reapply pressure (f), joining the melted ends of the pipes. After the joint has cooled, the pipe is removed from the Proweld tool. Butt fusion welding tools are shown in Figs. 3-6 and 3-7.

Butt fusion (welding) has many advantages over the socket fusion bonding process. Both pipe and pipe fittings can be joined by this process. The butt fusion tool shown in Fig. 3-6 is made in two models to handle pipe up to 4 in. in size.

In the actual welding process the butt ends of pipe, pipe and fittings, or fittings alone are clamped securely in perfect alignment, in preparation for the fusion process. Heat is applied through the heating mirror at temperatures of up to 500°F. A temperature-indicating light shows when the desired temperature is below the set point and turns off automatically when the desired temperature is reached.

A portable repair gun, Fig. 3-8, is made for minor repairs of thermoplastic pipes or tanks. This unit can be used for back welding PVC, CPVC, PP, and PVDF joints.

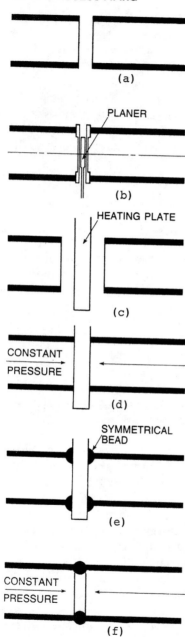

Fig. 3-4. Step-by-step butt fusion procedure. (Courtesy Asahi/America, Inc.)

Fig. 3-5. **Heating mirror used in butt fusion process.** *(Courtesy Asahi/America, Inc.)*

Fig. 3-6. **A portable butt fusion welding unit.** *(Courtesy Asahi/America, Inc.)*

Thermoplastic piping, fittings, and valves possess another unique characteristic, high insulating qualities. Therefore, under many conditions, no insulation is required on thermoplastic piping, fittings, and valves. Typical methods of anchoring and guiding plastic piping is shown in Figs. 3-9, 3-10, and 3-11.

Fig. 3-7.   A butt fusion tool assembled. *(Courtesy Asahi/America, Inc.)*

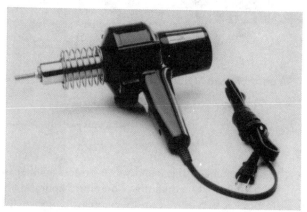

Fig. 3-8.   A repair gun for minor repairs of thermoplastics. *(Courtesy Asahi/America, Inc.)*

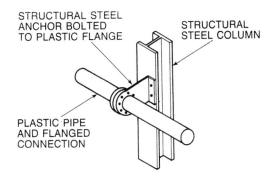

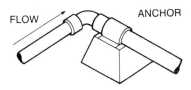

A TYPICAL METHOD OF ANCHORAGE

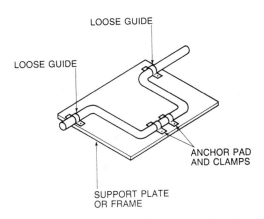

Fig. 3-9. Typical plastic piping anchor methods. *(Courtesy Asahi/ America, Inc.)*

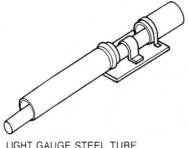

LIGHT GAUGE STEEL TUBE

ANNULAR ENDS TO BE
RADIUSED OR FLARED

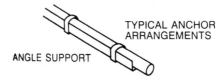

TYPICAL ANCHOR
ARRANGEMENTS

ANGLE SUPPORT

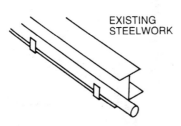

EXISTING
STEELWORK

Fig. 3-10. Typical plastic piping anchor methods. *(Courtesy Asahi/ America, Inc.)*

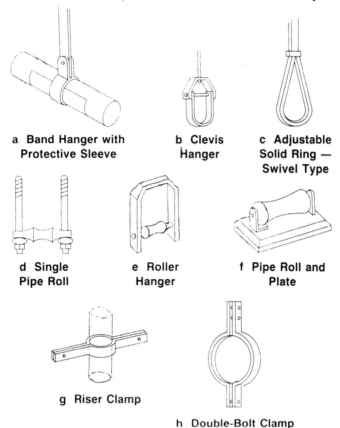

a Band Hanger with Protective Sleeve

b Clevis Hanger

c Adjustable Solid Ring — Swivel Type

d Single Pipe Roll

e Roller Hanger

f Pipe Roll and Plate

g Riser Clamp

h Double-Bolt Clamp

Fig. 3-11. Typical plastic piping hangers. *(Courtesy Asahi/America, Inc.)*

Glossary of Terms Used in the Plastics Industry

Abrasion Resistance—Ability to withstand the effects of repeated wearing, rubbing, scraping, etc.

Acids—One of a class of substances compounded of hydrogen and one or more other elements, capable of uniting with a base to form a salt and, in aqueous solution, turning blue litmus paper red.

Acrylate Resins—A class of thermoplastic resins produced by polymerization of acrylic acid derivatives.

Adhesive—A substance capable of holding materials together by surface attachment.

Aliphatic—Derived from or related to fats and other derivatives of the paraffin hydrocarbons, including unsaturated compounds of the ethylene and acetylene series.

Alkalies—Compounds capable of neutralizing acids and usually characterized by an acrid taste. Can be mild like baking soda or highly caustic like lye.

Alkyd Resins—A class of resins produced by condensation of a polybasic acid or anhydride and a polyhydric alcohol.

Allyl Resins—A class of resins produced from an ester or other derivative of allyl alcohol by polymerization.

Aromatic—A large class of cyclic organic compounds derived from, or characterized by, the presence of the benzene ring and its homologs.

Bond—To attach by means of an adhesive.

Cast Resin—A resinous product prepared by pouring liquid resins into a mold and heat-treating the mass to harden it.

Catalysis—The acceleration (or retardation) of the speed of a chemical reaction by the presence of a comparatively small amount of a foreign substance called a catalyst.

Cellulose Acetate—A class of resins made from a cellulose base, either cotton linters or purified wood pulp, by the action of acetic anhydride and acetic acid.

Compound—A combination of ingredients before being processed or made into a finished product. Sometimes used as a synonym for material formulation.

Copolymer—The product of simultaneous polymerization of two or more polymerizable chemicals, commonly known as monomers.

Crazing—Fine cracks at or under the surface of a plastic.

Creep—The unit elongation of a particular dimension under load for a specific time following the initial elastic elongation caused by load application. It is expressed usually in inches per unit of time.

Dielectric Constant—Specific inductive capacity. The dielectric constant of a material is the ratio of the capacitance of a condenser having that material as dielectric, to the capacity of the same condenser having a vacuum as dielectric.

Elastomer—The name applied to substances having rubber-like properties.

Electrical Properties—Primarily the resistance of a plastic to the passage of electricity.

Elongation—The capacity to take deformation before failure in tension, expressed as a percentage of the original length.

Ester—A compound formed by the elimination of waste during the reaction between an alcohol and an acid. Many esters are liquids. They are frequently used as plasticizers in rubber and plastic compounds.

Extender—A material added to a plastic composition to reduce its cost.

Extrusion—Method of processing plastic in a continuous or extended form by forcing heat-softened plastic through an opening shaped like the cross-section of the finished product.

Fuse—To join two plastic parts by softening the material by heat or solvents.

Generic—Common names for types of plastic materials. They may be either chemical terms or coined names. They contrast with trademarks which are the property of one company.

Impact Strength—Resistance or mechanical energy absorbed by a plastic part to such shocks as dropping and hard blows.

Impermeability—Permitting no passage into or through a material.

Injection Molding—Method of forming a plastic to the desired shape by forcing heat-softened plastic into a relatively cool cavity where it rapidly solidifies.

Ketones—Compounds containing the carbonyl group (CO) to which is attached two alkyl groups. Ketones, such as methyl ethyl ketone, are commonly used as solvents for resins and plastics.

Lubricant—A substance used to decrease the friction between solid faces, and sometimes used to improve processing characteristics of plastic compositions.

Moisture Resistance—Ability to resist absorption of water.

Monomer—The simplest repeating structural unit of a polymer; for addition polymers this represents the original unpolymerized compound.

Non-Toxic—Non-poisonous.

Organic Chemical—Originally applied to chemicals derived from living organisms, as distinguished from "inorganic" chemicals found in minerals and inanimate substances; modern chemists define organic chemicals more exactly as those which contain the element carbon.

Phenolic Resins—Resins made by reaction of a phenolic compound or tar acid with an aldehyde; more commonly applied to thermosetting resins made from pure phenol.

Plastic—A material that contains as an essential ingredient an organic substance of large molecular weight, is solid in its finished state, and, at some stage in its manufacture or in its processing into finished articles, can be shaped by flow.

Plasticity—A property of plastics and resins which allows the material to be deformed continuously and permanently without rupture upon the application of a force that exceeds the yield value of the material.

Plasticizer—A liquid or solid incorporated in natural and synthetic resins and related substances to develop such properties as resiliency, elasticity, and flexibility.

Polyethylenes—A class of resins formed by polymerizing ethylene, a gas obtained from petroleum hydrocarbons.

Polymer—A product resulting from a chemical change involving the successive addition of a large number of relatively small molecules (monomers) to form the polymer, and whose molecular weight is usually a multiple of that of the original substance.

Polymerization—Chemical change resulting in the formation of a new compound whose molecular weight is usually a multiple of that of the original substance.

Polyvinyl Chloride (PVC)—Polymerized vinyl chloride, a synthetic resin, which when plasticized or softened with other chemicals has some rubber-like properties. It is derived from acetylene and anhydrous hydrochloric acid.

Resin—An organic substance, generally synthetic, which is used as a base material for the manufacture of some plastics.

Rigid Plastic—A plastic which has a stiffness or apparent modulus of elasticity greater than 100,000 psi at 23°C when determined in accordance with the Standard Method of Test for Stiffness in Flexure of Plastics.

Solvent—The medium within which a substance is dissolved; most commonly applied to liquids used to bring particular solids into solution; e.g., acetone is a solvent for PVC.

Specific Gravity—Ratio of the mass of a body to the mass of an equal volume of water at 4°C, or some other specified temperature.

Specific Heat—Ratio of the thermal capacity of a substance to that of water at 15°C.

Stabilizer—A chemical substance which is frequently added to plastic compounds to inhibit undesirable changes in the material, such as discoloration due to heat or light.

Strength—The mechanical properties of a plastic, such as a load- or weight-carrying ability, and ability to withstand sharp blows. Strength properties include tensile, flexural, and tear strength, toughness, flexibility, etc.

Thermal Conductivity—Capacity of a plastic material to conduct heat.

Thermal Expansion—The increase in length of a dimension under the influence of a change in temperature.

Thermoplastic Materials—Materials which soften when heated to normal processing temperatures without the occurrence of appreciable chemical change, but are quickly hardened by cooling. Unlike the thermosetting materials they can be reheated to soften, and recooled to "set," almost indefinitely; they may be formed and reformed many times by heat and pressure.

Thermosetting—Plastic materials which undergo a chemical change and harden permanently when heated in processing. Further heating will not soften these materials.

Vinyl Chloride Plastics—Plastics based on resins made by the polymerization of vinyl chloride or copolymerization of vinyl chloride with minor amounts (not over 50 percent) of other unsaturated compounds.

Vinyl Plastics—Plastics based on resins made from vinyl monomers, except those specifically covered by other classifications, such as acrylic and styrene plastics. Typical vinyl plastics are polyvinyl chloride, polyvinyl acetate, polyvinyl alcohol, and polyvinyl butyral, and copolymers of vinyl monomers with unsaturated compounds.

Water Absorption—The percentages by weight of water absorbed by a sample immersed in water. Dependent upon area exposed.

Welding—The joining of two or more pieces of plastic by fusion of the material in the pieces at adjoining or nearby areas either with or without the addition of plastic from another source.

This list of terms used in the plastic industry will be helpful in understanding the use of plastics in the piping trade, such as PVC, CPVC, PP, and PVDF.

CHAPTER 4

Mathematics

Plumbing and heating are important factors in construction, and are vital to the health, morale, and welfare of building occupants. Piping is the basic material in plumbing, and can be compared to the veins and arteries of the human body. Plumbing pipe is classified as supply and waste, with the supply piping providing pure water to fixtures and waste piping removing that liquid.

Design is important in planning piping systems. Piping must be durable, have leak-proof joints, be of proper size for the intended purpose, and be in accordance with codes set up in almost all localities for purposes of protecting the health and welfare of the individual and public. In order to plan and compute well, a knowledge of mathematics is of great importance to the plumber and steam fitter. Therefore, a general knowledge of mathematics and access to certain reference tables are essential.

Symbols

The various processes in mathematics are usually indicated by symbols for convenience and brevity. The following are symbols commonly used.

$=$ means equal to, or equality;

$-$ means minus, less, or subtraction;

$+$ means plus, or addition;

$\times$ means multiplied by, or multiplication;

$\div$ or / means divided by, or division;

^{2} are indexes or powers, meaning that the number
^{3} to which they are added is to be squared or cubed; thus, 2^2 means 2 squared; 2^3 means 2 cubed;

$\left.\begin{array}{l} : \text{ is to} \\ :: \text{ so is} \\ : \text{ to} \end{array}\right\}$ are signs of proportion;

$\sqrt{}$ is the radical sign and means that the square root of the number before which it is placed is to be extracted;

$\sqrt[3]{}$ means that the cube root of the number before which it is placed is to be extracted;

—— the bar indicates that all of the numbers under it are to be taken together;

( ) the parentheses mean that all of the numbers between are to be taken as one quantity;

. the decimal point means decimal parts; thus 2.5 means $2\frac{5}{10}$, 0.46 means $\frac{46}{100}$;

$^\circ$ means degrees;

$: : :$ means hence;

π means ratio of the circumference of a circle to its diameter; numerically 3.1416;

$''$ means inches, seconds, or second;

$'$ means feet, minutes, or prime.

Abbreviations

In addition to the symbols just given, certain abbreviations and definitions are used. The practice of writing "pounds per square inch" instead of "lbs. per sq. in." is not preferred because in reading it the eye has to travel faster, resulting in fatigue and less speed in reading. The same thing is true of the excessive use of capital letters. It is a psychological fact that the omission of these capital letters results in less fatigue to the reader, though he may not be conscious of the fact.

The following abbreviations are commonly used:

A or a = area
A.W.G. = American wire gauge
B & S = Brown and Sharpe wire gauge (American wire gauge)
B or b = breadth
bbl. = barrels
bhp = brake horse power
B.M. = board measure
Btu = British thermal unit(s)
B.W.G. = Birmingham wire gauge
C of g = center of gravity
cond. = condensing
cu. = cubic
cyl. = cylinder
D or d = depth or diameter
deg. = degree(s)
diam. = diameter
evap. = evaporation
F = coefficient of friction; Fahrenheit
F or f = force or factor of safety
ft. lbs. = foot-pounds
gals. = gallons
H or h = height, or head of water
HP = horsepower
IHP = indicated horsepower
L or l = length
lbs. = pounds
lbs. per sq. in. = pounds per square inch
o.d. = outside diameter (pipes)
oz. = ounce(s)
P or p = pressure or load
psi = pounds per square inch
pt. = pint
R or r = radius
rpm = revolutions per minute
□′ = square feet
sq. ft. = square foot (feet)
sq. in. = square inch(es)
T or t = thickness, or temperature
temp. = temperature
V or v = velocity

vol. = volume
W or w = weight
W.I. = wrought iron

Definitions

Abstract Number—One that does not refer to any particular object.
Acute Triangle—One that has three acute angles.
Altitude (of a parallelogram or trapezoid)—The perpendicular distance between its parallel sides.
Altitude (of a prism)—The perpendicular distance between its bases.
Altitude (of a triangle)—A line drawn perpendicular to the base from the angle opposite.
Analysis—The process of investigating principles and solving problems independently of set rules.
Angle—The difference in direction of two lines proceeding from the same point called the vertex.
Area—The surface included within the lines that bound a plane figure.
Arithmetic—The science of numbers and the art of computation.
Base (of a triangle)—The side on which it may be supposed to stand.
Board Measure—A unit for measuring lumber being a volume of a board 12 in. wide, 1 ft. long, and 1 in. thick.
Circle—A plane figure bounded by a curved line, called the circumference, every point of which is equally distant from a point within, called the center.
Complex Fraction—One whose numerator or denominator is a fraction.
Composite Numbers—A number that can be divided by other integers besides itself and one.
Compound Fraction—A fraction of a fraction.
Compound Numbers—Units of two or more denominations of the same kind.
Concrete Numbers—A number used to designate objects or quantities.
Cone—A body having a circular base, and whose convex surface tapers uniformly to the vertex.
Cube—A parallelopipedon whose faces are equal squares.
Cubic Measure—A measure of volume involving three dimensions—length, breadth, and thickness.

Cylinder—A body bounded by a uniformly curved surface, its ends being equal and parallel circles.

Decimal Scale—One in which the order of progression is uniformly ten.

Demonstration—Process of reasoning by which a truth or principle is established.

Denomination—Name of the unit of a concrete number.

Diagonal (of a plane figure)—A straight line joining the vertices of two angles not adjacent.

Diameter (of a circle)—A line passing through its center and terminated at both ends by the circumference.

Diameter (of a sphere)—A straight line passing through the center of the sphere, and terminated at both ends by its surface.

Equilateral Triangle—One that has all its sides equal.

Even Number—A number that can be exactly divided by two.

Exact Divisor of a Number—A whole number that will divide that number without a remainder.

Factors—One of two or more quantities that, when multiplied together, produces a given quantity.

Factors of a Number—Numbers that, when multiplied together, make that number.

Fraction—A number that expresses equal parts of a whole thing or quantity.

Frustum (of a pyramid or cone)—The part that remains after cutting off the top by a plane parallel to the base.

Geometry—The branch of pure mathematics that treats of space and its relations.

Greatest Common Divisor—The greatest number that will exactly divide two or more numbers.

Hypotenuse (of a right triangle)—The side opposite the right angle.

Improper Fraction—One whose numerator equals or exceeds its denominator.

Integer—A number that represents whole things.

Involution—The multiplication of a quantity by itself any number of times; raising a number to a given power.

Isosceles Triangle—One that has two of its sides equal.

Least Common Multiple—Least number that is exactly divisible by two or more numbers.

Like Numbers—Same kind of unit, expressing the same kind of quantity.

Mathematics—The science of quantity.

Measure—That by which the extent, quantity, capacity, volume, or dimensions in general is ascertained by some fixed standard.

Mensuration—The process of measuring.

Multiple of a Number—Any number exactly divisible by that number.

Number—A unit or collection of units.

Obtuse Triangle—One that has one obtuse angle.

Odd Number—A number that cannot be divided by two.

Parallelogram—Quadrilateral that has its opposite sides parallel.

Parallelopipedon—A prism bounded by six parallelograms, the opposite ones being parallel and equal.

Percentage—Rate per hundred.

Perimeter (of a polygon)—The sum of its sides.

Perpendicular (of a right triangle)—The side that forms a right angle with the base.

Plane Figure—A plane surface.

Polygon—A plane figure bounded by straight lines.

Power—Product arising from multiplying a number.

Prime Factor—A prime number used as a factor.

Prime Number—A number exactly divisible by some number other than one or itself.

Prism—A solid whose ends are equal and parallel polygons, and whose sides are parallelograms.

Problem—A question requiring an operation.

Proper Fraction—One whose numerator is less than its denominator.

Pyramid—A body having for its base a polygon, and for its other sides or facets three or more triangles that terminate in a common point called the vertex.

Quadrilateral—A plane figure bounded by four straight lines and having four angles.

Quantity—That which can be increased, diminished, or measured.

Radius (of a circle)—A line extending from its center to any point on the circumference. It is one-half the diameter.

Radius (of a sphere)—A straight line drawn from the center to any point on the surface.

Rectangle—A parallelogram with all its angles right angles.

Rhomboid—A parallelogram whose opposite sides only are equal, but whose angles are not right angles.

Rhombus—A parallelogram whose sides are all equal, but whose angles are not right angles.

Right Triangle—One that has a right angle.

Root—A factor repeated to produce a power.

Rule—A prescribed method of performing an operation.

Scale—Order of progression on which any system of notation is founded.

Scalene Triangle—One that has all of its sides unequal.

Simple Fraction—One whose numerator and denominator are whole numbers.

Simple Number—Either an abstract number or a concrete number of but one denomination.

Slant Height (of a cone)—A straight line from the vertex to the circumference of the base.

Slant Height (of a pyramid)—The perpendicular distance from its vertex to one of the sides of the base.

Sphere—A body bounded by a uniformly curved surface, all the points of which are equally distant from a point within called the center.

Square—A rectangle whose sides are equal.

Trapezium—A quadrilateral having no two sides parallel.

Trapezoid—A quadrilateral, two of whose sides are parallel and two oblique.

Triangle—A plane figure bounded by three sides, and having three angles.

Uniform Scale—One in which the order of progression is the same throughout the entire succession of units.

Unit—A single thing or a definite quantity.

Unity—Unit of an abstract number.

Unlike Numbers—Different kinds of units, used to express different kinds of quantity.

Varying Scale—One in which the order of progression is not the same throughout the entire succession of units.

Notation and Numeration

By definition, *notation* in arithmetic is *the writing down of figures to express a number*. A numeration is *the reading of the number or collection of figures already written*.

By means of the ten figures that follow, any number can be expressed.

0 1 2 3 4 5 6 7 8 9

Table 4-1A. Numeration Table

Names of Units	Billions			Millions			Thousands			Units			Thousandths			
Grouping of the Units	Hundred-billions	Ten-billions	Billions	Hundred-millions	Ten-millions	Millions	Hundred-thousands	Ten-thousands	Thousands	Hundreds	Tens	Units	Decimal point	Tenths	Hundredths	Thousandths
	7	8	6,	5	4	3,	2	0	1,	2	8	2,	.	4	8	9

The system in Table 4-1A is called *Arabic notation,* and it is the system in ordinary everyday use.

Table 4-1B shows the *Roman* system of notation often used, especially to denote the year of construction or manufacture.

The following ten formulas include the elementary operations of arithmetic.

1. The sum = all the parts added.
2. The difference = the minuend − the subtrahend.
3. The minuend = the subtrahend + the difference.
4. The subtrahend = the minuend − the difference.
5. The product = the multiplicand × the multiplier.
6. The multiplicand = the product ÷ the multiplier.
7. The multiplier = the product ÷ the multiplicand.
8. The quotient = the dividend ÷ the divisor.
9. The dividend = the quotient × the divisor.
10. The divisor = the dividend ÷ the quotient.

Table 4-1B. Roman Numerals

I = 1	VIII = 8	XV = 15	XL = 40	D =500
II = 2	IX = 9	XVI = 16	L = 50	M =1000
III = 3	X = 10	XVII = 17	LX = 60	$\overline{X}$ =10,000
IV = 4	XI = 11	XVIII = 18	LXX = 70	$\overline{M}$ =1,000,000
V = 5	XII = 12	XIX = 19	LXXX = 80	
VI = 6	XIII = 13	XX = 20	XC = 90	
VII = 7	XIV = 14	XXX = 30	C = 100	

Addition

The sign of addition is + and is read *plus:* thus 7 + 3 is read *seven plus three.*

> **Rule A.** Write the numbers to be added so that like orders of units stand in the same column.
> **Rule B.** Beginning with the lowest order, or at the right hand, add each column separately, and if the sum can be expressed by one figure, write it under the column added.
> **Rule C.** If the sum of any column contains more than one figure, write the unit figure under the column added, and add the remaining figure or figures to the next column.

Examples:

7,060	248,124	13,579,802
9,420	4,321	93
1,743	889,966	478,652
4,004	457,902	87,547,289
22,227 Ans.		

Use great care in placing the numbers in vertical lines, as irregularity in writing them down is one cause of mistakes.

Subtraction

The sign of subtraction is − and is read *minus;* thus 10 − 7 is read *ten minus seven* or *seven from ten.*

> **Rule A.** Write down the sum so that the units stand under the units, the tens under the tens, etc.
> **Rule B.** Begin with the units, and take the lower figure from the upper figure and put the remainder beneath the line.
> **Rule C.** If the lower figure is the larger, add ten to the upper figure, and then subtract and put the remainder down; this borrowed ten must be deducted from the next column of figures where it is represented by 1.

Examples:

892	2,572	9,999
46	1,586	8,971
846 remainder		

Multiplication

The sign of multiplication is × and is read *times* or *multiplied by;* thus 6 × 8 is read 6 times 8, or 6 multiplied by 8. The principle of multiplication is the same as addition; thus 3 × 8 = 24 is the same as 8 + 8 + 8 = 24.

Rule. Place the unit figure of the multiplier under the unit figure of the multiplicand and proceed as in the following examples:

Example—Multiply 846 by 8; and 478,692 by 143. Arrange them thus:

```
 846             487692
   8                143
6768            1463076
                1950768
                487692
                69739956
```

Rule. If the multiplier has ciphers at its end, place it as in the following examples:

Example—Multiply 83567 by 50; and 898 by 2800.

```
  83567          898
     50         2800
4178350       718400
              1796
              2514400
```

Division

The sign of division is ÷ and is read *divided by;* thus 8 ÷ 2 is read *eight divided by two.* There are two methods of division known as *short division* and *long division.*

Short Division
To divide by any number up to 12.

Rule. Put the dividend down with the divisor to the left of it, with a small curved line separating it, as in the following:

Example—Divide 7,865,432 by 6.

$$6)\underline{7,865,432}$$
$$1,310,905 \text{ —} \tfrac{2}{6} \text{ or } .3$$

Here at the last, 6 into 32 goes 5 times and 2 over; always place the number that is left over as a fraction. This would be $\tfrac{2}{6}$, the top figure being the remainder and the bottom figure the divisor. It should be put close to the quotient; thus $1,310,905\tfrac{2}{6}$.

To divide by any number up to 12 with a cipher or ciphers after it, as 20, 70, 500, 7000, etc.

Rule. Place the sum down as in the last example, then mark off from the right of the dividend as many figures as there are ciphers in the divisor; also mark off the ciphers in the divisor; then divide the remaining figures by the number remaining in the divisor, thus:

Example—Divide 9,876,804 by 40.

$$40)\underline{9,876,804.}$$
$$246,920.1$$

Long Division
To divide any number by a large divisor of two or more figures.

Example—Divide 18,149 by 56.

$$56)\ \underline{18,149}(324\tfrac{5}{56}$$
$$\underline{168}$$
$$134$$
$$\underline{112}$$
$$229$$
$$\underline{224}$$
$$5$$

In the above operation, the process is as follows: As neither 1 nor 18 will contain the divisor, take the first three figures (181) for the first partial dividend. The number 56 is contained in 181 three times, with a remainder. Write the 3 as the first figure in the quotient, and then multiply the divisor by this quotient figure thus: 3 times 56 is 168, which when subtracted from 181 leaves 13. To this remainder annex bring down 4, the next figure in the dividend, thus forming 134 which is the next partial dividend. The number 56 is contained in 134 two times with a remainder. Thus, 2 times 56 is 112, which subtracted from 134 leaves 22. To the remainder bring down 9, the last figure in the dividend, forming 229, the last partial dividend. The number 56 is contained in 229 four times with a remainder. Thus: 4 times 56 is 224, which subtracted from 229 gives 5, the final remainder, thus completing the operation of long division.

Factors

Numbers 4 and 5 are factors of 20, because 4 multiplied by 5 equals 20.

> **Rule.** Divide the given number by any prime factor; divide the quotient in the same manner, and so continue the division until the quotient is a prime number. The several divisors and the last quotient will be the prime factors required.

Example—What are the prime factors of 798?

$$
\begin{array}{r|r}
2 & \underline{798} \\
3 & \underline{399} \\
7 & \underline{133} \\
19 & \underline{19}
\end{array}
$$

Greatest Common Divisor

Number 5 is the greatest common divisor of 10 and 15, because it is the greatest number that will exactly divide each of them.

> **Rule.** Write the numbers in a line, with a vertical line at the left, and divide by any factor common to all the numbers. Divide the quotient in like manner, and continue dividing till a set of quotients is obtained that are prime to each other. Multiply all the divisors together and the product will be the greatest common divisor sought.

Example—What is the greatest common divisor of 72, 120, and 440?

4	72	120	440
2	18	30	110
	9	15	55

Least Common Divisor

Number 6 is the least common multiple of 2 and 3, because it is the least number exactly divisible by those numbers.

Rule. Resolve the given numbers into their prime factors. Multiply together all the prime factors of the largest number, and such prime factors of the other numbers as are not found in the largest number. Their product will be the least common multiple. When a prime factor is repeated in any of the given numbers, it must be taken as many times in the multiple as the greatest number of times it appears in any of the given numbers.

Example—Find the least common multiple of 60, 84, and 132.

$$60 = 2 \times 2 \times 3 \times 5$$
$$84 = 2 \times 2 \times 3 \times 7$$
$$132 = 2 \times 2 \times 3 \times 11$$
$$(2 \times 2 \times 3 \times 11) \times 5 \times 7 = 4620$$

Fractions

If a unit or whole number is divided into two equal parts, one of these parts is called *one-half*, written $\frac{1}{2}$.

To reduce a common fraction to its lowest terms:
Rule. Divide both terms by their greatest common divisor.

Example:

$$\frac{9}{15} = \frac{3}{5}$$

To change an improper fraction to a mixed number:

Rule. Divide the numerator by the denominator; the quotient is the whole number, and the remainder placed over the denominator is the fraction.

Example:

$$\frac{23}{4} = 5\frac{3}{4}$$

To change a mixed number to an improper fraction:

Rule. Multiply the whole number by the denominator of the fraction; to the product add the numerator and place the sum over the denominator.

Example:

$$1\frac{3}{8} = \frac{11}{8}$$

To reduce a compound to a single fraction, and to multiply fractions:

Rule. Multiply the numerators together for a new numerator and the denominators together for a new denominator.

Example:

$$\frac{1}{2} \text{ of } \frac{2}{3} = \frac{2}{6}; \text{ also } \frac{1}{2} \times \frac{2}{3} = \frac{2}{6}$$

To reduce a complex fraction to a simple fraction:

Rule. The numerator and denominator must each first be given the form of a simple fraction; then multiply the numerator of the upper fraction by the denominator of the lower for the new numerator, and the denominator of the upper by the numerator of the lower for the new denominator.

Example:

$$\frac{\dfrac{7}{8}}{1\dfrac{3}{4}} = \frac{\dfrac{7}{8}}{\dfrac{7}{4}} = \frac{28}{56} = \frac{1}{2}$$

To add fractions:
Rule. Reduce them to a common denominator, add the numerators, and place their sum over the common denominator.

Example:

$$\frac{1}{2} + \frac{1}{4} = \frac{4 + 2}{8} = \frac{6}{8} = \frac{3}{4}$$

To subtract fractions:
Rule. Reduce them to a common denominator, subtract the numerators, and place the difference over the common denominator.

Example:

$$\frac{1}{2} - \frac{1}{4} = \frac{4 - 2}{8} = \frac{2}{8} = \frac{1}{4}$$

To multiply fractions:
Rule. (Multiplying by a whole number.) Multiply the numerator or divide the denominator by the whole number.

Example:

$$\frac{1}{2} \times 3 = \frac{3}{2} = 1\frac{1}{2}$$

To divide fractions:
Rule. (Dividing by a whole number.) Divide the numerator, or multiply the denominator by the whole number.

Example:

$$(\text{dividing})\frac{10}{13} \div 5 = \frac{2}{13}; (\text{multiplying})\frac{10}{13} \div 5 = \frac{10}{65} = \frac{2}{13}$$

Rule. (Dividing by a fraction.) Invert the divisor and proceed as in multiplication.

Example:

$$\frac{3}{4} \div \frac{5}{7} = \frac{3}{4} \times \frac{7}{5} = \frac{21}{20} = 1\frac{1}{20}$$

Table 4-2. Numeration of Decimals

Ten thousands	Thousands	Hundreds	Tens	Units	Decimal point	Tenths	Hundredths	Thousandths	Ten thousandths	Hundred thousandths
1	2	3	4	5	.	1	2	3	4	5
5th order	4th order	3rd order	2nd order	1st order		1st order	2nd order	3rd order	4th order	5th order

Integers   *Decimals*

Decimal Fractions

Any decimal or combination of a decimal and integer may be read by applying Table 4-2.

The important thing about decimals is to always plainly put down the decimal point. In case of a column of figures, as in addition, care should be taken to have all the decimal points exactly under each other.

> *To reduce a decimal to a common fraction:*
> **Rule.** Write down the denominator and reduce the common fraction thus obtained to its lowest terms.

Example:

$$.25 = \frac{25}{100} = \frac{1}{4}$$

To add and subtract decimals:

Rule. Place the numbers in a column with the decimal points under each other and proceed as in simple addition or subtraction.

Examples:

Addition	Subtraction
.5	1.25
.25	.75
1.75	.50
2.50	

To multiply decimals:

Rule. Proceed as in simple multiplication and point off as many places as there are places in the multiplier and multiplicand.

Example:

$$.1 \times .0025 = .00025$$

To divide decimals:

Rule. Proceed as in simple division, and from the right hand of the quotient, point off as many places for decimals as the decimal places in the dividend exceed those in the divisor.

Example:

$$1.5 \div .25 = 6$$

To reduce common fractions to decimals:

Rule. Divide the numerator by the denominator and carry out the division to as many decimal places as desired.

Example:

$$\frac{4}{5} = 4 \div 5 = .8$$

Table 4-3. Fractions and Decimal Equivalents

$1/64$ = .015625	$11/32$ = .34375	$43/64$ = .671875
$1/32$ = .03125	$23/64$ = .359375	$11/16$ = .6875
$3/64$ = .046875	$3/8$ = .375	$45/64$ = .703125
$1/16$ = .0625	$25/64$ = .390625	$23/32$ = .71875
$5/64$ = .078125	$13/32$ = .40625	$47/64$ = .734375
$3/32$ = .09375	$27/64$ = .421875	$3/4$ = .75
$7/64$ = .109375	$7/16$ = .4375	$49/64$ = .765625
$1/8$ = .125	$29/64$ = .453125	$25/32$ = .78125
$9/64$ = .140625	$15/32$ = .46875	$51/64$ = .796875
$5/32$ = .15625	$31/64$ = .484375	$13/16$ = .8125
$11/64$ = .171875	$1/2$ = .5	$53/64$ = .828125
$3/16$ = .1875	$33/64$ = .515625	$27/32$ = .84375
$13/64$ = .203125	$17/32$ = .53125	$55/64$ = .859375
$7/32$ = .21875	$35/64$ = .546875	$7/8$ = .875
$15/16$ = .234375	$9/16$ = .5625	$57/64$ = .890625
$1/4$ = .25	$37/64$ = .578125	$29/32$ = .90625
$17/64$ = .265625	$19/32$ = .59375	$59/64$ = .921875
$9/32$ = .28125	$39/64$ = .609375	$15/16$ = .9375
$19/64$ = .296875	$5/8$ = .625	$61/64$ = .953125
$5/16$ = .3125	$41/64$ = .640625	$31/32$ = .96875
$21/64$ = .328125	$21/32$ = .65625	$63/64$ = .984375

The decimal equivalents of common fractions given in Table 4-3 will be found very useful.

Ratio and Proportion

A ratio is virtually a fraction. When two ratios are equal, the four terms form a proportion. Thus 2:4: :3:6, which is read as 2 is to 4 as 3 is to 6. Sometimes the = sign is placed between the two ratios instead of the sign : :, thus 2:4 = 3:6.

Rule. Two quantities of *different* kinds cannot form the terms of a ratio.

Rule. The product of the extremes equals the product of the means.

Example:

4:8 = 2:4, or 4 × 4 = 8 × 2, or 16 = 16

Rule of Three. When three terms of a proportion are given, the method of finding the fourth term is called the *rule of three.*

Example—If five boxes of nails cost $16, what will 25 boxes cost? Let X equal the unknown term; then

5 boxes : 25 boxes = \$16 : \$X. 5 × X

$$= 25 \times 16 \ X = \frac{25 \times 16}{5} = \$80$$

Percentage

A profit of 6% means a gain of \$6 on every \$100. Note carefully with respect to the symbol %: 5% means 5/100 which, when reduced to a decimal (as is necessary in making a calculation), becomes .05. However, .05% has a quite different value; thus, 0.05% means .05/100 which, when reduced to a decimal, becomes .0005; that is, 5/100 of 1%.

Rule. If the decimal has more than two places, the figures that follow the hundredths place signify parts of 1%.

Example—If the list price of screws is \$16 per 1000, what is the net cost with 5% discount for cash?

$$5\% = \frac{5}{100} = .05; 16 \times .05 = 80¢; \$16 - 80¢ = \$15.20$$

Powers of Numbers

The *square* of a number is its second power; the *cube*, its third power. Thus,

the square of 2 = 2 × 2 = 4; the cube of 2
$$= 2 \times 2 \times 2 = 8$$

The power to which a number is raised is indicated by a small *superior* figure called an *exponent*. Thus,

$$2^2 = 2 \times 2 = 4; 2^3 = 2 \times 2 \times 2 = 8$$

Roots of Numbers (Evolution)

In the equation 2 × 2 = 4, the number 2 is the root for which the power (4) is produced. The radical sign $\sqrt{}$ placed over a number means the root of the number is to be extracted. Thus $\sqrt{4}$ means that the square root of 4 is to be extracted. The *index* of the root is a small figure placed over the radical.

Rule. (Square root.) As shown in the example, point off the given number into groups of two places each, beginning with units. If there are decimals, point these off likewise, beginning at the decimal point and supplying as many ciphers as may be needed. Find the greatest number whose square is less than the first left-hand group, and place it as the first figure in the quotient. Subtract its square from the left-hand group, and annex the two figures of the second group to the remainder for a dividend. Double the first figure of the quotient for a partial divisor; find how many times the latter is contained in the dividend, exclusive of the right-hand figure in the quotient, and annex it to the right of the partial divisor, forming the complete divisor. Multiply this divisor by the second figure in the quotient, and subtract the product from the dividend. To the remainder, bring down the next group and proceed as before, in each case doubling the figures in the root already found to obtain the trial divisor. Should the product of the second figure in the root by the completed divisor be greater than the dividend, erase the second figure both from the quotient and from the divisor, and substitute the next smaller figure, or one small enough to make the product of the second figure by the divisor less than or equal to the dividend.

Example:

$$
\begin{array}{r|l}
\multicolumn{2}{c}{3.'14'15'92'65'36'\underline{\,|\,1.77245+}} \\
& 1 \\
27 & 214 \\
& \underline{189} \\
347 & 2515 \\
& \underline{2429} \\
3542 & 8692 \\
& \underline{7084} \\
35444 & 160865 \\
& \underline{141776} \\
354485 & 1908936 \\
& \underline{1772425}
\end{array}
$$

Rule. (Cube root.) As shown in the example, separate the number into groups of three figures each, beginning at the units. Find the greatest cube in the left-hand group and write its root for the first figure of the required root. Cube this root, subtract the result from the left-hand group, and to the remainder annex the next group for a dividend. For a partial divisor, take three times the square of the root already found (considered as tens), and divide the dividend by it. The quotient (or the quotient diminished) will be the second figure of the root. To this partial divisor add three times the product of the first figure on the root (considered as tens) by the second figure, and also the square of the second figure. This sum will be the complete divisor. Multiply the complete divisor by the second figure of the root, subtract the product from the dividend, and to the remainder annex the next group for a new dividend. Proceed in this manner until all the groups have been annexed. The result will be the cube root required. Table 4-4 can be a great help in determining the square, cube, square root, or cube root of numbers up to 100.

Example:

$$1',881',365',963',625' \qquad 12345$$

$$
\begin{array}{rl}
300 \times 1^2 & = 300 \,\overline{|\,881} \\
30 \times 1 \quad \times 2 & = \ 60 \\
2^2 & = \ \ \underline{4} \\
& \quad 364 \ \underline{|\,728} \\
& \qquad \ \ \ 153365
\end{array}
$$

$$
\begin{array}{rl}
300 \times 12^2 & = 43200 \\
30 \times 12 \quad \times 3 & = 1080 \\
3^2 & = \ \ \ \ \underline{9} \\
& \ 44289 \ |\,132867 \\
& \qquad \quad \ 20498963
\end{array}
$$

$$
\begin{array}{rl}
300 \times 123^2 & = 4538700 \\
30 \times 123 \quad \times 4 & = \ \ 14760 \\
4^2 & = \ \ \ \ \ \ \underline{16} \\
& 4553476 \ \ |\,18213904 \\
& \qquad \qquad \ 2285059625
\end{array}
$$

$$
\begin{array}{rl}
300 \times 1234^2 & = 456826800 \\
30 \times 1234 \times 5 & = \ \ 185100 \\
5^2 & = \ \ \ \ \ \ \ \underline{25} \\
& 457011925 \ \ |\,\underline{2285059625}
\end{array}
$$

Table 4-4. Squares, Cubes, Square Roots, and Cube Roots

No.	Square	Cube	Square Root	Cube Root	Reciprocal
51	2601	132651	7.14142	3.70842	0.01960
52	2704	140608	7.21110	3.73251	0.01923
53	2809	148877	7.28010	3.75628	0.01886
54	2916	157464	7.34846	3.77976	0.01851
55	3025	166375	7.41619	3.80295	0.01818
56	3136	175616	7.48331	3.82586	0.01785
57	3249	185193	7.54983	3.84850	0.01754
58	3364	195112	7.61577	3.87087	0.01724
59	3481	205379	7.68114	3.89299	0.01694
60	3600	216000	7.74596	3.91486	0.01666
61	3721	226981	7.81024	3.93649	0.01639
62	3844	238328	7.87400	3.95789	0.01612
63	3969	250047	7.93725	3.97905	0.01587
64	4096	262144	8.00000	4.00000	0.01562
65	4225	274625	8.06225	4.02072	0.01538
66	4356	287496	8.12403	4.04124	0.01515
67	4489	300763	8.18535	4.06154	0.01492
68	4624	314432	8.24621	4.08165	0.01470
69	4761	328509	8.30662	4.10156	0.01449
70	4900	343000	8.36660	4.12128	0.01428
71	5041	357911	8.42614	4.14081	0.01408
72	5184	373248	8.48528	4.16016	0.01388
73	5329	389017	8.54400	4.17933	0.01369
74	5476	405224	8.60232	4.19833	0.01351
75	5625	421875	8.66025	4.21716	0.01333
76	5776	438976	8.71779	4.23582	0.01315
77	5929	456533	8.77496	4.25432	0.01298
78	6084	474552	8.83176	4.27265	0.01282
79	6241	493039	8.88819	4.29084	0.01265
80	6400	512000	8.94427	4.30886	0.01250
81	6561	531441	9.00000	4.32674	0.01234
82	6724	551368	9.05538	4.34448	0.01219
83	6889	571787	9.11043	4.36207	0.01204
84	7056	592704	9.16515	4.37951	0.01190
85	7225	614125	9.21954	4.39682	0.01176
86	7396	636056	9.27361	4.41400	0.01162
87	7569	658503	9.32737	4.43104	0.01149
88	7744	681472	9.38083	4.44796	0.01136
89	7921	704969	9.43398	4.46474	0.01123
90	8100	729000	9.48683	4.48140	0.01111
91	8281	753571	9.53939	4.49794	0.01098
92	8464	778688	9.59166	4.51435	0.01086
93	8649	804357	9.64365	4.53065	0.01075
94	8836	830584	9.69535	4.54683	0.01063
95	9025	857375	9.74679	4.56290	0.01052
96	9216	884736	9.79795	4.57885	0.01041
97	9409	912673	9.84885	4.59470	0.01030
98	9604	941192	9.89949	4.61043	0.01020
99	9801	970299	9.94987	4.62606	0.01010
100	10000	1000000	10.00000	4.64158	0.01000

Table 4-4. Squares, Cubes, Square Roots, and Cube Roots (Cont'd)

No.	Square	Cube	Square Root	Cube Root	Reciprocal
51	2601	132651	7.14142	3.70842	0.01960
52	2704	140608	7.21110	3.73251	0.01923
53	2809	148877	7.28010	3.75628	0.01886
54	2916	157464	7.34846	3.77976	0.01851
55	3025	166375	7.41619	3.80295	0.01818
56	3136	175616	7.48331	3.82586	0.01785
57	3249	185193	7.54983	3.84850	0.01754
58	3364	195112	7.61577	3.87087	0.01724
59	3481	205379	7.68114	3.89299	0.01694
60	3600	216000	7.74596	3.91486	0.01666
61	3721	226981	7.81024	3.93649	0.01639
62	3844	238328	7.87400	3.95789	0.01612
63	3969	250047	7.93725	3.97905	0.01587
64	4096	262144	8.00000	4.00000	0.01562
65	4225	274625	8.06225	4.02072	0.01538
66	4356	287496	8.12403	4.04124	0.01515
67	4489	300763	8.18535	4.06154	0.01492
68	4624	314432	8.24621	4.08165	0.01470
69	4761	328509	8.30662	4.10156	0.01449
70	4900	343000	8.36660	4.12128	0.01428
71	5041	357911	8.42614	4.14081	0.01408
72	5184	373248	8.48528	4.16016	0.01388
73	5329	389017	8.54400	4.17933	0.01369
74	5476	405224	8.60232	4.19833	0.01351
75	5625	421875	8.66025	4.21716	0.01333
76	5776	438976	8.71779	4.23582	0.01315
77	5929	456533	8.77496	4.25432	0.01298
78	6084	474552	8.83176	4.27265	0.01282
79	6241	493039	8.88819	4.29084	0.01265
80	6400	512000	8.94427	4.30886	0.01250
81	6561	531441	9.00000	4.32674	0.01234
82	6724	551368	9.05538	4.34448	0.01219
83	6889	571787	9.11043	4.36207	0.01204
84	7056	592704	9.16515	4.37951	0.01190
85	7225	614125	9.21954	4.39682	0.01176
86	7396	636056	9.27361	4.41400	0.01162
87	7569	658503	9.32737	4.43104	0.01149
88	7744	681472	9.38083	4.44796	0.01136
89	7921	704969	9.43398	4.46474	0.01123
90	8100	729000	9.48683	4.48140	0.01111
91	8281	753571	9.53939	4.49794	0.01098
92	8464	778688	9.59166	4.51435	0.01086
93	8649	804357	9.64365	4.53065	0.01075
94	8836	830584	9.69535	4.54683	0.01063
95	9025	857375	9.74679	4.56290	0.01052
96	9216	884736	9.79795	4.57885	0.01041
97	9409	912673	9.84885	4.59470	0.01030
98	9604	941192	9.89949	4.61043	0.01020
99	9801	970299	9.94987	4.62606	0.01010
100	10000	1000000	10.00000	4.64158	0.01000

Rule. (Roots higher than the cube.) The fourth root is the square root of the square root; the sixth root is the cube root of the square root, or the square root of the cube root. Other roots are most conveniently found by the use of logarithms.

The Metric System

The important feature of the metric system is that it is based upon the *decimal scale.* Thus, a knowledge of decimals is needed before taking up this system.

The metric system is the decimal system of measures and weights, with the meter and the gram as the bases. The unit of length (the meter) was intended to be and is very nearly one ten-millionth part of the distance measured on a meridian from the equator to the pole, or 39.37079 in. The other primary units of measure such as the *square meter*, the *cubic meter*, the *liter*, and the *gram* are based on the meter.

Following is the *metric* system of weights and measures. Table 4-5 shows the conversion of millimeters into inches, and Table 4-6 shows the conversion of inches into millimeters.

Milli expresses the 1000th part.
Centi expresses the 100th part.
Deci expresses the 10th part.
Deka expresses 10 times the value.
Hecto expresses 100 times the value.
Kilo expresses 1000 times the value.

Length

1 mm.	= 1 millimeter	= 1/1000 of a meter	=	.03937	in.
10 mm.	= 1 centimeter	= 1/100 of a meter	=	.3937	in.
10 cm.	= 1 decimeter	= 1/10 of a meter	=	3.937	in.
10 dm.	= 1 meter	= 1 meter	=	39.37	in.
10 m.	= 1 dekameter	= 10 meters	=	32.8	ft.
10 dm.	= 1 hectometer	= 100 meters	=	328.09	ft.
10 hm.	= 1 kilometer	= 1000 meters	=	.62137	mile

Table 4-5. Millimeters to Inches

mm.	inches		mm.	inches		mm.	inches
$1/50$ = 0.00079			$26/50$ = 0.02047			2 = 0.07874	
$2/50$ = 0.00157			$27/50$ = 0.02126			3 = 0.11811	
$3/50$ = 0.00236			$28/50$ = 0.02205			4 = 0.15748	
$4/50$ = 0.00315			$29/50$ = 0.02283			5 = 0.19685	
$5/50$ = 0.00394			$30/50$ = 0.02362			6 = 0.23622	
$6/50$ = 0.00472			$31/50$ = 0.02441			7 = 0.27559	
$7/50$ = 0.00551			$32/50$ = 0.02520			8 = 0.31496	
$8/50$ = 0.00630			$33/50$ = 0.02598			9 = 0.35433	
$9/50$ = 0.00709			$34/50$ = 0.02677			10 = 0.39370	
$10/50$ = 0.00787			$35/50$ = 0.02756			11 = 0.43307	
$11/50$ = 0.00866			$36/50$ = 0.02835			12 = 0.47244	
$12/50$ = 0.00945			$37/50$ = 0.02913			13 = 0.51181	
$13/50$ = 0.01024			$38/50$ = 0.02992			14 = 0.55118	
$14/50$ = 0.01102			$39/50$ = 0.03071			15 = 0.59055	
$15/50$ = 0.01181			$40/50$ = 0.03150			16 = 0.62992	
$16/50$ = 0.01260			$41/50$ = 0.03228			17 = 0.66929	
$17/50$ = 0.01339			$42/50$ = 0.03307			18 = 0.70866	
$18/50$ = 0.01417			$43/50$ = 0.03386			19 = 0.74803	
$19/50$ = 0.01496			$44/50$ = 0.03465			20 = 0.78740	
$20/50$ = 0.01575			$45/50$ = 0.03543			21 = 0.82677	
$21/50$ = 0.01654			$46/50$ = 0.03622			22 = 0.86614	
$22/50$ = 0.01732			$47/50$ = 0.03701			23 = 0.90551	
$23/50$ = 0.01811			$48/50$ = 0.03780			24 = 0.94488	
$24/50$ = 0.01890			$49/50$ = 0.03858			25 = 0.98425	
$25/50$ = 0.01969			1 = 0.03937			26 = 1.02362	

Square Measure

1 sq. centimeter = 0.1550 sq. in.
1 sq. in. = 6.452 sq. centimeters
1 sq. decimeter = 0.1076 sq. ft.
1 sq. ft. = 9.2903 sq. decimeters
1 sq. meter = 1.196 sq. yd.
1 sq. yd. = 0.8361 sq. meter
1 acre = 3.954 sq. rods
1 sq. rod = 0.2529 acre
1 hectare = 2.47 acres
1 acre = 0.4047 hectare
1 sq. kilometer = 0.386 sq. mile
1 sq. mile = 2.59 sq. kilometers

Table 4-6. Inches to Millimeters

In.	0	$^1/_{16}$	$^1/_8$	$^3/_{16}$	$^7/_{16}$	$^1/_4$	$^5/_{16}$	$^3/_8$
0	0.0	1.6	3.2	4.8	6.4	7.9	9.5	11.1
1	25.4	27.0	28.6	30.2	31.7	33.3	34.9	36.5
2	50.8	52.4	54.0	55.6	57.1	58.7	60.3	61.9
3	76.2	77.8	79.4	81.0	82.5	84.1	85.7	87.3
4	101.6	103.2	104.8	106.4	108.0	109.5	111.1	112.7
5	127.0	128.6	130.2	131.8	133.4	134.9	136.5	138.1
6	152.4	154.0	155.6	157.2	158.8	160.3	161.9	163.5
7	177.8	179.4	181.0	182.6	184.2	185.7	187.3	188.9
8	203.2	204.8	206.4	208.0	209.6	211.1	212.7	214.3
9	228.6	230.2	231.8	233.4	235.0	236.5	238.1	239.7
10	254.0	255.6	257.2	258.8	260.4	261.9	263.5	265.1
11	279.4	281.0	282.6	284.2	285.7	287.3	288.9	290.5
12	304.8	306.4	308.0	309.6	311.1	312.7	314.3	315.9
13	330.2	331.8	333.4	335.0	336.5	338.1	339.7	341.3
14	355.6	357.2	358.8	360.4	361.9	363.5	365.1	366.7
15	381.0	382.6	384.2	385.8	387.3	388.9	390.5	392.1
16	406.4	408.0	409.6	411.2	412.7	414.3	415.9	417.5
17	431.8	433.4	435.0	436.6	438.1	439.7	441.3	442.9
18	457.2	458.8	460.4	462.0	463.5	465.1	466.7	468.3
19	482.6	484.2	485.8	487.4	488.9	490.5	492.1	493.7
20	508.0	509.6	511.2	512.8	514.3	515.9	517.5	519.1
21	533.4	535.0	536.6	538.2	539.7	541.3	542.9	544.5
22	558.8	560.4	562.0	563.6	565.1	566.7	568.3	569.9
23	584.2	585.8	587.4	589.0	590.5	592.1	593.7	595.3

In.	$^1/_2$	$^9/_{16}$	$^5/_8$	$^{11}/_{16}$	$^3/_4$	$^{13}/_{16}$	$^7/_8$	$^{15}/_{16}$
0	12.7	14.3	15.9	17.5	19.1	20.6	22.2	23.8
1	38.1	39.7	41.3	42.9	44.4	46.0	47.6	49.2
2	63.5	65.1	66.7	68.3	69.8	71.4	73.0	74.6
3	88.9	90.5	92.1	93.7	95.2	96.8	98.4	100.0
4	114.3	115.9	117.5	119.1	120.7	122.2	123.8	125.4
5	139.7	141.3	142.9	144.5	146.1	147.6	149.2	150.8
6	165.1	166.7	168.3	169.9	171.5	173.0	174.6	176.2
7	190.5	192.1	193.7	195.3	196.9	198.4	200.0	201.6
8	215.9	217.5	219.1	220.7	222.3	223.8	225.4	227.0
9	241.3	242.9	244.5	246.1	247.7	249.2	250.8	252.4
10	266.7	268.3	269.9	271.5	273.1	274.6	276.2	277.8
11	292.1	293.7	295.3	296.9	298.4	300.0	301.6	303.2
12	317.5	319.1	320.7	322.3	323.8	325.4	327.0	328.6
13	342.9	344.5	346.1	347.7	349.2	350.8	352.4	354.0
14	368.3	369.9	371.5	373.1	374.6	376.2	377.8	379.4
15	393.7	395.3	396.9	398.5	400.0	401.6	403.2	404.8
16	419.1	420.7	422.3	423.9	425.4	427.0	428.6	430.2
17	444.5	446.1	447.7	449.3	450.8	452.4	454.0	455.6
18	469.9	471.5	473.1	474.7	476.2	477.8	479.4	481.0
19	495.3	496.9	498.5	500.1	501.6	503.2	504.8	506.4
20	520.7	522.3	523.9	525.5	527.0	528.6	530.2	531.8
21	546.1	547.7	549.3	550.9	552.4	554.0	555.6	557.2
22	571.5	573.1	574.7	576.3	577.8	579.4	581.0	582.6
23	596.9	598.5	600.1	601.7	603.2	604.8	606.4	608.0

Table Weights

1 gram = 0.0527 ounce   1 ounce = 28.35 grams
1 kilogram = 2.2046 lbs.   1 lb. = 0.4536 kilogram
1 metric ton = 1.1023 English   1 English ton = 0.9072 metric
ton   ton

Approximate Metric Equivalents

1 decimeter = 4 inches   1 liter = 1.06 qt. liquid; 0.9
1 meter = 1.1 yards   qt. dry
1 kilometer = $\frac{5}{8}$ mile   1 hectoliter = $2\frac{5}{8}$ bushel
1 hectare = $2\frac{1}{2}$ acres   1 kilogram = $2\frac{1}{5}$ lbs.
1 stere or cu. meter = $\frac{1}{4}$ 1 metric ton = 2200 lbs.
cord

Long Measure

12 inches (in. or ")   = 1 foot (ft. or ')
3 feet   = 1 yard (yd.)
$5\frac{1}{2}$ yards or $16\frac{1}{2}$ feet   = 1 rod (rd.)
40 rods   = 1 furlong (fur.)
8 furlongs or 320 rods = 1 statute mile (mi.)

Nautical Measure

6080.26 ft. or 1.15156 statute miles = 1 nautical mile
3 nautical miles = 1 league
60 nautical miles or 69.168 statute miles = 1 degree (at the
equator)
360 degrees = circumference of
earth at equator

Square Measure

144 square inches (sq. in.) = 1 square foot (sq. ft.)

9 sq. ft. = 1 square yard (sq. yd.)
$30\frac{1}{4}$ sq. yd. = 1 square rod or perch (sq. rd. or P.)
640 acres = 1 square mile (sq. mi.)

Mensuration

Mensuration is the process of measuring objects that occupy space; for instance, finding the length of a line, the area of a triangle, or the volume of a cube.

Triangles

Figures bounded by three sides are called triangles; there are numerous kinds due to varying angles and length of sides.

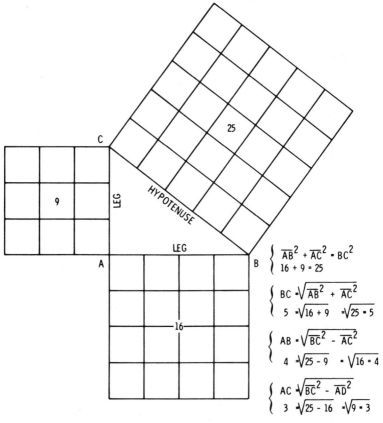

$$\begin{cases} \overline{AB}^2 + \overline{AC}^2 = BC^2 \\ 16 + 9 = 25 \end{cases}$$

$$\begin{cases} BC = \sqrt{\overline{AB}^2 + \overline{AC}^2} \\ 5 = \sqrt{16 + 9} = \sqrt{25} = 5 \end{cases}$$

$$\begin{cases} AB = \sqrt{\overline{BC}^2 - \overline{AC}^2} \\ 4 = \sqrt{25 - 9} = \sqrt{16} = 4 \end{cases}$$

$$\begin{cases} AC = \sqrt{\overline{BC}^2 - \overline{AD}^2} \\ 3 = \sqrt{25 - 16} = \sqrt{9} = 3 \end{cases}$$

Fig. 4-1.   Right triangle showing mathematical relationships.

To find the length of the hypotenuse of a right triangle:
Rule. The hypotenuse is equal to the square root of the sum of the squares of each leg, as shown in Fig. 4-1.

To find the length of either leg of a right triangle:
Rule. Either leg is equal to the square root of the difference between the square of the hypotenuse and the square of the other leg (Fig. 4-1).

To find the area of any triangle:
Rule. Multiply the base by half the perpendicular height. Thus, if the base is 12 ft. and the height 8 ft., the area $= \frac{1}{2}$ of $8 \times 12 = 48$ sq. ft.

Quadrilaterals

Any plane figure bounded by four sides is a quadrilateral, as shown in Fig. 4-2.

To find the area of a trapezium:
Rule. Join two of its opposite angles, and thus divide it into two triangles. Measure this line and call it the base of each triangle. Measure the perpendicular height of each triangle above the base line. Then find the area of each triangle by the previous rule; their sum is the area of the whole figure.

To find the area of a trapezoid:
Rule. Multiply half the sum of the two parallel sides by the perpendicular distance between them.

To find the area of a square:
Rule. Multiply the base by the height; that is, multiply the length by the breadth.

To find the area of a rectangle:
Rule. Multiply the length by the breadth.

To find the area of a parallelogram:
Rule. Multiply the base by the perpendicular height.

Polygons

These comprise the numerous figures having more than four sides, named according to the number of sides, thus:

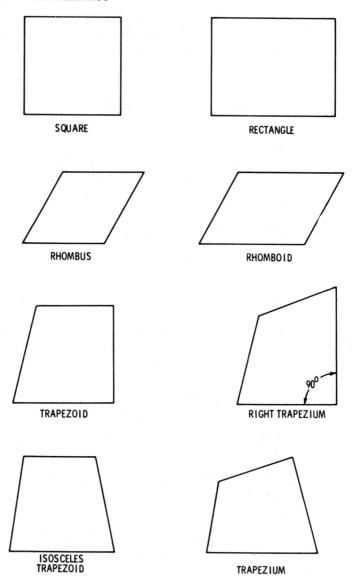

Fig. 4-2. Various quadrilaterals.

pentagon . . . . 5 sides
hexagon. . . . . 6 sides
heptagon . . . . 7 sides
octagon . . . . 8 sides
nonagon . . . . 9 sides
decagon. . . . . 10 sides

To find the area of a polygon:
Rule. Multiply the sum of the sides (perimeter of the polygon) by the perpendicular dropped from its center to one of its sides, and half the product will be the area. This rule applies to all regular polygons.

To find the area of any regular polygon when the length of a side only is given:
Rule. Multiply the square of the sides by the figure for "Area when side = 1" opposite the polygon in Table 4-7.

Table 4-7. Table of Regular Polygons

Number of sides	3	4	5	6	7	8	9	10	11	12
Area when side = 1	.433	1.	1.721	2.598	3.634	4.828	6.181	7.694	9.366	11.196

The Circle

The Greek letter π (called pi) is used to represent 3.1416, the circumference of a circle whose diameter is 1. The circumference of a circle equals the diameter multiplied by 3.1416. The reason why the decimal .7854 is used to calculate the area of a circle is explained in Fig. 4-3.

To find the circumference of a circle:
Rule. Multiply the diameter by 3.1416.

To find the diameter of a circle (circumference given):
Rule. Divide the circumference by 3.1416.

To find the area of a circle:
Rule. Multiply the square of the diameter by .7854. (See Fig. 4-3.)

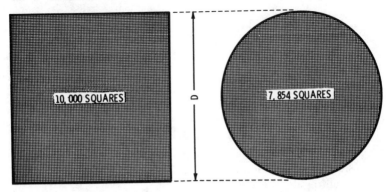

Fig. 4-3. Diagram illustrating why the decimal .7854 is used to find the area of a circle.

To find the diameter of a circle (area given):
Rule. Extract the square root of the area divided by .7854.

To find the area of a sector of a circle:
Rule. Multiply the arc of the sector by half the radius.

To find the area of a segment of a circle:
Rule. Find the area of the sector which has the same arc and also the area of the triangle formed by the radii and chord; take the sum of these areas if the segment is greater than 1800; take the difference if less.

To find the area of a ring:
Rule. Take the difference between the areas of the two circles.

To find the area of an ellipse:
Rule. Multiply the product of the two diameters by .7854.

The following list shows the relation of a circle to an equal, inscribed, and circumscribed square.

Diameter of circle	× .88623	= side of equal square
Circumference of circle	× .28209	
Circumference of circle	× 1.1284	= perimeter of equal square

Diameter of circle	× .7071	= side of inscribed square
Circumference of circle	× .22508	
Area of circle	× .90031 ÷ diameter	

Area of circle	× 1.2732	= area of circumscribed square
Area of circle	× .63662	= area of inscribed square
Side of square	× 1.4142	= diameter of circumscribed circle
Side of square	× 4.4428	= circumference
Side of square	× 1.1284	= diameter of equal circle
Side of square	× 3.5449	= circumference of equal circle
Perimeter of square	× .88623	= circumference of equal circle
Square inches	× 1.2732	= circular inches

Solids

Finding the volume of the solids involves the multiplication of three dimensions—length, breadth, and thickness.

To find the volume of a solid:
Rule. Multiply the area of the base by the perpendicular height.

To find the volume of a rectangular solid:
Rule. Multiply the length, breadth, and height.

To find the surface of a cylinder:
Rule. Multiply 3.1416 by the diameter times the length.

To find the volume of a cylinder:
Rule. Multiply .7854 by the diameter squared of the base times the length of the cylinder.

To find the surface of a sphere:
Rule. Multiply the area of its great circle by 4.

To find the volume of a sphere:
Rule. Multiply .7854 by the cube of the diameter, and then take $\frac{2}{3}$ of the product.

To find the volume of a segment of a sphere:
Rule. To three times the square of the radius of the segment's base, add the square of the depth or height; then multiply this sum by the depth, and the product by .5236.

To find the surface of a cylindrical ring:
Rule. To the thickness of the ring, add the inner diameter; multiply this sum by the thickness, and the product again by 9.8696.

To find the volume of a cylindrical ring:
Rule. To the thickness of the ring, add the inner diameter; multiply this sum by the square of the thickness, and the product again by 2.4674.

To find the slant area of a cone:
Rule. Multiply 3.1416 by the diameter of the base and by one-half the slant height.

To find the slant area of the frustum of a cone:
Rule. Multiply half the slant height by the sum of the circumferences.

To find the volume of a cone:
Rule. Multiply the area of the base by the perpendicular height, and by $\frac{1}{3}$.

To find the volume of a frustum of a cone:
Rule. Find the sum of the squares of the two diameters (d, D), and add to this the product of the two diameters multiplied by .7854, and by one-third the height (h).

To find the volume of a pyramid:
Rule. Multiply the area of the base by one-third of the perpendicular height.

To find the volume of a rectangular solid:
Rule. Multiply the length, breadth, and thickness.

To find the volume of a rectangular wedge:
Rule. Find the area of one of the triangle ends and multiply by the distance between ends.

Mensuration of Surfaces and Volumes

Area of rectangle = length × breadth.
Area of triangle = base × $\frac{1}{2}$ perpendicular height.
Diameter of circle = radius × 2.
Circumference of circle = diameter × 3.1416.
Area of circle = square of diameter × .7854.
Area of sector of circle =

$$\frac{\text{area of circle} \times \text{number of degrees in arc}}{360}$$

Area of surface of cylinder = circumference × length + area of two ends.

To find diameter of circle having given area: Divide the area by .7854, and extract the square root.

To find the volume of a cylinder: Multiply the area of the section in square inches by the length in inches = the volume in cubic inches. Cubic inches divided by 1728 = volume in cubic feet.

Surface of a sphere = square of diameter × 3.1416.
Volume of a sphere = cube of diameter × .5236.
Side of an inscribed cube = radius of a sphere × 1.1547.

Area of the base of a pyramid or cone, whether round, square, or triangular, multiplied by one-third of its height = the volume.

Diameter × .8862 = side of an equal square.
Diameter × .7071 = side of an inscribed square.
Radius × 6.2832 = circumference.
Circumference = 3.5446 × $\sqrt{\text{Area of circle}}$.
Diameter = 1.1283 × $\sqrt{\text{Area of circle}}$.
Length of arc = No. of degrees × .017453 radius.
Degrees in arc whose length equals radius = 57° 2958′.
Length of an arc of 1° = radius × .017453.
Length of an arc of 1 min. = radius × .0002909.
Length of an arc of 1 sec. = radius × .0000048.

π = Proportion of circumference to diameter = 3.1415926.

π^2 = 9.8696044.

$\sqrt{\pi}$ = 1.7724538.

$\text{Log}\pi$ = 0.49715.

$\dfrac{1}{\pi}$ = 0.31831.

$1/360$ = .002778.

$\dfrac{360}{\pi}$ = 114.59.

Lineal feet	× .00019	= Miles.
Lineal yards	× .0006	= Miles.
Square inches	× .007	= Square feet.
Square feet	× .111	= Square yards.
Square yards	× .0002067	= Acres.
Acres	× 4840.	= Square yards.
Cubic inches	× .00058	= Cubic feet.
Cubic feet	× .03704	= Cubic yards.
Circular inches	× .00546	= Square feet.
Cyl. inches	× .0004546	= Cubic feet.
Cyl. feet	× .02909	= Cubic yards.
Links	× .22	= Yards.
Links	× .66	= Feet.
Feet	× 1.5	= Links.
Width in chains	× 8	= Acres per mile.
Cubic feet	× 7.48	= U.S. gallons.
Cubic inches	× .004329	= U.S. gallons.
U.S. gallons	× .13367	= Cubic feet.
U.S. gallons	× 231	= Cubic inches.
Cubic feet	× .8036	= U.S. bushel.
Cubic inches	× .000466	= U.S. bushel.
Lbs. Avoir	× .009	= Cwt. (112)
Lbs. Avoir	× .00045	= Tons (2240)
Cubic feet of water	× 62.5	= Lbs. Avoir.
Cubic inch of water	× .03617	= Lbs. Avoir.
13.44 U.S. gallons of water		= 1 cwt.
268.8 U.S. gallons of water		= 1 ton.
1.8 cubic feet of water		= 1 cwt.
35.88 cubic feet of water		= 1 ton.
Column of water, 12 inches high, and 1 inch in diameter		= .341 lbs.
U.S. bushel	× .0495	= Cubic yards.
U.S. bushel	× 1.2446	= Cubic feet.
U.S. bushel	× 150.42	= Cubic inches.

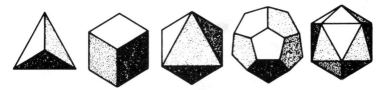

Fig. 4-4.   The five regular solids.

Table 4-8.   Surfaces and Volumes of Regular Solids

Number of Sides	Name	Area Edge =	Contents Edge = 1
4	Tetrahedron	1.7320	0.1178
6	Hexahedron	6.0000	1.0000
8	Octahedron....................	3.4641	0.4714
12	Dodecahedron	20.6458	7.6631
20	Icosahedron	8.6603	2.1817

To find the volume of irregular solids:

Rule. Divide the irregular solid into different figures; and the sum of their volumes, found by the preceding problems, will be the volume required. If the figure is a compound solid, whose two ends are equal plane figures, the volume may be found by multiplying the area of one end by the length. To find the volume of a piece of wood or stone that is craggy or uneven, put it into a tub or cistern, and pour in as much water as will just cover it; then take it out and find the contents of that part of the vessel through which the water has descended and it will be the volume required.

To find the surface and volume of any of the five regular solids shown in Fig. 4-4:

Rule. (Surface.) Multiply the area given in Table 4-8 by the square of the edge of the solid.

Rule. (Volume.) Multiply the contents given in Table 4-8 by the cube of the given edge.

Trigonometric Functions

Every triangle has only six parts—3 sides and 3 angles. When any three of these parts are given (provided one of them is a side), the

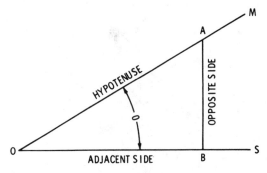

Fig. 4-5. A triangle, AOB, for expressing trigonometric functions as ratios.

other parts may be determined. Fig. 4-5 illustrates the parts considered in expressing trigonometric functions. It will be noted in this triangle that angle ABO = 90°. In this triangle the trigonometric functions, expressed as ratios, are as follows:

$$\text{\textit{Sine} of the angle} = \frac{AB}{AO} = \frac{\text{opposite side}}{\text{hypotenuse}}$$

$$\text{\textit{Cosine} of the angle} = \frac{OB}{OA} = \frac{\text{adjacent side}}{\text{hypotenuse}}$$

$$\text{\textit{Tangent} of the angle} = \frac{AB}{OB} = \frac{\text{opposite side}}{\text{adjacent side}}$$

$$\text{\textit{Cotangent} of the angle} = \frac{OB}{AB} = \frac{\text{adjacent side}}{\text{opposite side}}$$

$$\text{\textit{Secant} of the angle} = \frac{OA}{OB} = \frac{\text{hypotenuse}}{\text{adjacent side}}$$

$$\text{\textit{Cosecant} of the angle} = \frac{OA}{AB} = \frac{\text{hypotenuse}}{\text{opposite side}}$$

Natural Functions

These are virtually ratios, but by taking what corresponds to the hypotenuse OA, in the triangle AOB in Fig. 4-5, as a radius of unity length of a circle, the denominators of the ratios are unity or 1. These denominators disappear, leaving only the numerators; that is, a line instead of a ratio or function. These lines are the so-called *natural functions*. Thus, in Fig. 4-6:

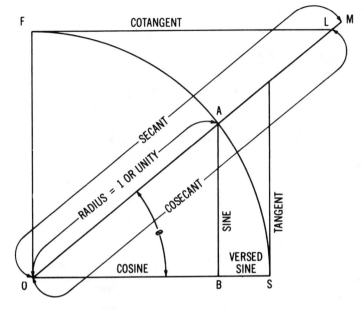

Fig. 4-6.   Natural trigonometric functions.

$$Sine \text{ angle} = \frac{AB}{radius} = \frac{AB}{1} = AB$$

$$Cosine \text{ angle} = \frac{radius}{OB} = OB$$

$$Tangent \text{ angle} = \frac{MS}{OS} = \frac{MS}{radius} = MS$$

$$Cotangent \text{ angle} = tangent \text{ of complement of angle}$$

$$= \frac{OM}{OF} = \frac{OM}{radius} = OM$$

$$Secant \text{ angle} = \frac{OM}{OS} = \frac{OM}{radius} = OM$$

$$Cosecant \text{ angle} = secant \text{ of complement angle} =$$

$$\frac{OL}{OF} = \frac{OL}{radius} = OL$$

The natural trigonometric functions shown in Fig. 4-7 are the ones of value in ordinary calculations and should be thoroughly understood. They are used in connection with Table 4-9, as illustrated by the following example.

Table 4-9. Natural Trigonometric Functions

Degs.	Sine	Cosine	Tangent	Secant	Degs.	Sine	Cosine	Tangent	Secant
0	0.00000	1.0000	0.00000	1.0000	46	0.7193	0.6947	1.0355	1.4395
1	0.01745	0.9998	0.01745	1.0001	47	0.7314	0.6820	1.0724	1.4663
2	0.03490	0.9994	0.03492	1.0006	48	0.7431	0.6691	1.1106	1.4945
3	0.05234	0.9986	0.05241	1.0014	49	0.7547	0.6561	1.1504	1.5242
4	0.06976	0.9976	0.06993	1.0024	50	0.7660	0.6428	1.1918	1.5557
5	0.08716	0.9962	0.08749	1.0038	51	0.7771	0.6293	1.2349	1.5890
6	0.10453	0.9945	0.10510	1.0055	52	0.7880	0.6157	1.2799	1.6243
7	0.12187	0.9925	0.12278	1.0075	53	0.7986	0.6018	1.3270	1.6616
8	0.1392	0.9903	0.1405	1.0098	54	0.8090	0.5878	1.3764	1.7013
9	0.1564	0.9877	0.1584	1.0125	55	0.8192	0.5736	1.4281	1.7434
10	0.1736	0.9848	0.1763	1.0154	56	0.8290	0.5592	1.4826	1.7883
11	0.1908	0.9816	0.1944	1.0187	57	0.8387	0.5446	1.5399	1.8361
12	0.2079	0.9781	0.2126	1.0223	58	0.8480	0.5299	1.6003	1.8871
13	0.2250	0.9744	0.2309	1.0263	59	0.8572	0.5150	1.6643	1.9416
14	0.2419	0.9703	0.2493	1.0306	60	0.8660	0.5000	1.7321	2.0000
15	0.2588	0.9659	0.2679	1.0353	61	0.8746	0.4848	1.8040	2.0627
16	0.2756	0.9613	0.2867	1.0403	62	0.8829	0.4695	1.8807	2.1300
17	0.2924	0.9563	0.3057	1.0457	63	0.8910	0.4540	1.9626	2.2027
18	0.3090	0.9511	0.3249	1.0515	64	0.8988	0.4384	2.0503	2.2812
19	0.3256	0.9455	0.3443	1.0576	65	0.9063	0.4226	2.1445	2.3662
20	0.3420	0.9397	0.3640	1.0642	66	0.9135	0.4067	2.2460	2.4586
21	0.3584	0.9336	0.3839	1.0711	67	0.9205	0.3907	2.3559	2.5593
22	0.3746	0.9272	0.4040	1.0785	68	0.9272	0.3746	2.4751	2.6695
23	0.3907	0.9205	0.4245	1.0864	69	0.9336	0.3584	2.6051	2.7904
24	0.4067	0.9135	0.4452	1.0946	70	0.9397	0.3420	2.7475	2.9238
25	0.4226	0.9063	0.4663	1.1034	71	0.9455	0.3256	2.9042	3.0715
26	0.4384	0.8988	0.4877	1.1126	72	0.9511	0.3090	3.0777	3.2361
27	0.4540	0.8910	0.5095	1.1223	73	0.9563	0.2924	3.2709	3.4203
28	0.4695	0.8829	0.5317	1.1326	74	0.9613	0.2756	3.4874	3.6279
29	0.4848	0.8746	0.5543	1.1433	75	0.9659	0.2588	3.7321	3.8637
30	0.5000	0.8660	0.5774	1.1547	76	0.9703	0.2419	4.0108	4.1336
31	0.5150	0.8572	0.6009	1.1666	77	0.9744	0.2250	4.3315	4.4454
32	0.5299	0.8480	0.6249	1.1792	78	0.9781	0.2079	4.7046	4.8097
33	0.5446	0.8387	0.6494	1.1924	79	0.9816	0.1908	5.1446	5.2408
34	0.5592	0.8290	0.6745	1.2062	80	0.9848	0.1736	5.6713	5.7588
35	0.5736	0.8192	0.7002	1.2208	81	0.9877	0.1564	6.3138	6.3924
36	0.5878	0.8090	0.7265	1.2361	82	0.9903	0.1392	7.1154	7.1853
37	0.6018	0.7986	0.7536	1.2521	83	0.9925	0.12187	8.1443	8.2055
38	0.6157	0.7880	0.7813	1.2690	84	0.9945	0.10453	9.5144	9.5668
39	0.6293	0.7771	0.8098	1.2867	85	0.9962	0.08716	11.4301	11.474
40	0.6428	0.7660	0.8391	1.3054	86	0.9976	0.06976	14.3007	14.335
41	0.6561	0.7547	0.8693	1.3250	87	0.9986	0.05234	19.0811	19.107
42	0.6691	0.7431	0.9004	1.3456	88	0.9994	0.03490	28.6363	28.654
43	0.6820	0.7314	0.9325	1.3673	89	0.9998	0.01745	57.2900	57.299
44	0.6947	0.7193	0.9657	1.3902	90	1.0000	Inf.	Inf.	Inf.
45	0.7071	0.7071	1.0000	1.4142					

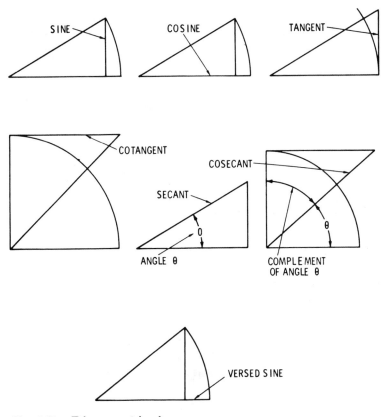

Fig. 4-7. Trigonometric shapes.

Example—In Fig. 4-8, two pipe lines 8 in. apart are to be connected with 30° elbows. What is the length of the offset OB and connecting pipe OA? From Table 4-9, tangent 60° = 1.73; length offset OB = 1.73 × 8 = 13.84. Again, from Table 4-9, secant 60° = 2; length connecting pipe OA = 8 × 2 = 16 in.

It is often necessary to figure the capacity of a round storage tank. The old standard methods involve several steps; the more steps used, the greater the possibility of error. Newer, simpler methods of computing tank capacities reduce the possibility of errors. A comparison of the two methods follows.

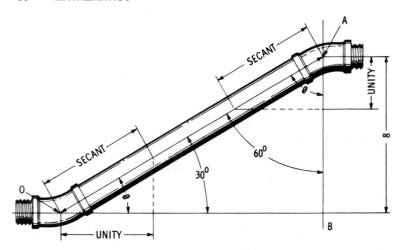

Fig. 4-8. **Two parallel pipe lines connected with 30° elbows illustrating the use of natural trigonometric functions in finding the offset and length of connecting pipes.**

Old method—tank measured in inches.

formula: $D^2 \times .7854 \times L \div 231 = $ gallons

Example—How many gallons are in a tank 12 in. in diameter and 60 in. long?

Solution:

$$
\begin{array}{cccc}
12 & 144 & 113.0976 & 29.3760 \\
\times\ 12 & \times\ .7854 & \times\quad\quad 60 & 231\overline{)6785.8560} \\
\hline
144 & 576 & 6785.8560 & \underline{462} \\
 & 720 & & 2165 \\
 & 1152 & & \underline{2079} \\
 & \underline{1008} & & 868 \\
 & 113.0976 & & \underline{693} \\
 & & & 1755 \\
 & & & \underline{1617} \\
 & & & 1386 \\
 & & & \underline{1386} \\
 & & & 0000 \\
\end{array}
$$

Answer: 29.3760 gallons in tank.

New method—tank measured in inches.

formula: $D^2 \times L \times .0034 =$ gallons

Example—How many gallons are in a tank 12 in. in diameter and 60 in. long?

Solution:

```
    12          144          8640
 ×  12        × 60        × .0034
   144         8640         34560
                            25920
                           29.3760
```

Answer: 29.3760 gallons in tank.

Old Method—tank measured in feet.

formula: $D^2 \times .7854 \times L \times 7.5 =$ gallons

Example—How many gallons are in a tank 1 ft. in diameter and 5 ft. long?

Solution:

```
    1        .7854        .7854        3.9270
 ×  1        × 1          × 5          × 7.5
    1        .7854        3.9270       19635
                                       27489
                                      29.4525
```

Answer: 29.4525 gallons.

Note: Using the old standard methods, there is a slight difference in the answers.

New method, using a combination of feet and inches.

formula: $D^2 \times L \times .0408$

Example—How many gallons are in a tank 12 in. in diameter and 5 ft. long?

Solution:

12	144	720
× 12	× 5	.0408
144	720	5760
		28800
		29.3760

Answer: 29.3760 gallons in tank.

Note: The simplicity of the new method is readily apparent, and the answers are virtually identical.

One of the most common problems in plumbers' mathematics is to figure the length of pipe when making a common offset in a piping run. Changes in direction using simple offsets are made by using elbows or, in some cases, wyes, where a cleanout fitting is desirable. The angle of the fitting is the number of degrees in the change of direction. Thus a 45° elbow changes the direction 45°. Fig. 4-9 shows the terms used for various parts of an offset. There are some simple formulas which can be used to calculate offsets. These formulas or multipliers are shown in Table 4-10.

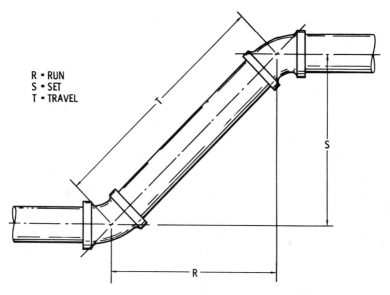

R = RUN
S = SET
T = TRAVEL

Fig. 4-9. Terms used for various parts of an offset.

Table 4-10. Constants for Measuring Offsets in Piping

known side	to find side	multiply side	using 5⅝ ell	using 11¼ ell	using 22½ ell	using 30 ell	using 45 ell	using 60 ell
S	T	S	10.19	5.13	2.61	2.00	1.41	1.15
S	R	S	10.16	5.03	2.41	1.73	1.00	.58
R	S	R	.10	.20	.41	.58	1.00	1.73
R	T	R	1.00	1.02	1.08	1.16	1.41	2.00
T	S	T	.10	.20	.38	.50	.71	.87
T	R	T	1.00	.98	.92	.87	.71	.50

Table 4-11. Areas of Circles

Diameter	Area	Diameter	Area	Diameter	Area	Diameter	Area
⅛	0.0123	10	78.54	30	706.86	65	3318.3
¼	0.0491	10½	86.59	31	754.77	66	3421.2
⅜	0.1105	11	95.03	32	804.25	67	3525.7
½	0.1964	11½	103.87	33	855.30	68	3631.7
⅝	0.3068	12	113.09	34	907.92	69	3739.3
¾	0.4418	12½	122.72	35	962.11	70	3848.5
⅞	0.6013	13	132.73	36	1017.9	71	3959.2
1	0.7854	13½	143.14	37	1075.2	72	4071.5
1⅛	0.9940	14	153.94	38	1134.1	73	4185.4
1¼	1.227	14½	165.13	39	1194.6	74	4300.8
1⅜	1.485	15	176.71	40	1256.6	75	4417.9
1½	1.767	15½	188.69	41	1320.3	76	4536.5
1⅝	2.079	16	201.06	42	1385.4	77	4656.6
1¾	2.405	16½	213.82	43	1452.2	78	4778.4
1⅞	2.761	17	226.98	44	1520.5	79	4901.7
2	3.142	17½	240.53	45	1590.4	80	5026.5
2¼	3.976	18	254.47	46	1661.9	81	5153.0
2½	4.909	18½	268.80	47	1734.9	82	5281.0
2¾	5.940	19	283.53	48	1809.6	83	5410.6
3	7.069	19½	298.65	49	1885.7	84	5541.8
3¼	8.296	20	314.16	50	1963.5	85	5674.5
3½	9.621	20½	330.06	51	2042.8	86	5808.8
3¾	11.045	21	346.36	52	2123.7	87	5944.7
4	12.566	21½	363.05	53	2206.2	88	6082.1
4½	15.904	22	380.13	54	2290.2	89	6221.1
5	19.635	22½	397.61	55	2375.8	90	6361.7
5½	23.758	23	415.48	56	2463.0	91	6503.9
6	28.274	23½	433.74	57	2551.8	92	6647.6
6½	33.183	24	452.39	58	2642.1	93	6792.9
7	38.485	24½	471.44	59	2734.0	94	6939.8
7½	44.179	25	490.87	60	2827.4	95	7088.2
8	50.265	26	530.93	61	2922.5	96	7238.2
8½	56.745	27	572.56	62	3019.1	97	7389.8
9	63.617	28	615.75	63	3117.2	98	7543.0
9½	70.882	29	660.52	64	3217.0	99	7697.7

Example—If the set, *S*, is 12 in. then to find the travel, *T*, using 45° elbows, Table 4-10 shows that the multiplier for 45° offsets is 1.41. Multiplying 12 × 1.41 shows that the travel, *T*, is 16.92 in. Rounding this off, *T* = 17.00 in. Table 4-10 is very useful when figuring offsets using standard fittings.

Table 4-12. Circumference of Circles

Diameter	Circumference	Diameter	Circumference	Diameter	Circumference	Diameter	Circumference
1/8	.3927	10	31.42	30	94.25	65	204.2
1/4	.7854	10½	32.99	31	97.39	66	207.3
3/8	1.178	11	34.56	32	100.5	67	210.5
1/2	1.571	11½	36.13	33	103.7	68	213.6
5/8	1.963	12	37.70	34	106.8	69	216.8
3/4	2.356	12½	39.27	35	110.0	70	219.9
7/8	2.749	13	40.84	36	113.1	71	223.0
1	3.142	13½	42.41	37	116.2	72	226.2
1⅛	3.534	14	43.98	38	119.4	73	229.3
1¼	3.927	14½	45.55	39	122.5	74	232.5
1⅜	4.320	15	47.12	40	125.7	75	235.6
1½	4.712	15½	48.69	41	128.8	76	238.8
1⅝	5.105	16	50.26	42	131.9	77	241.9
1¾	5.498	16½	51.84	43	135.1	78	245.0
1⅞	5.890	17	53.41	44	138.2	79	248.2
2	6.283	17½	54.98	45	141.4	80	251.3
2¼	7.069	18	56.55	46	144.5	81	254.5
2½	7.854	18½	58.12	47	147.7	82	257.6
2¾	8.639	19	59.69	48	150.8	83	260.8
3	9.425	19½	61.26	49	153.9	84	263.9
3¼	10.21	20	62.83	50	157.1	85	267.0
3½	11.00	20½	64.40	51	160.2	86	270.2
3¾	11.78	21	65.97	52	163.4	87	273.3
4	12.57	21½	67.54	53	166.5	88	276.5
4½	14.14	22	69.12	54	169.6	89	279.6
5	15.71	22½	70.69	55	172.8	90	282.7
5½	17.28	23	72.26	56	175.9	91	285.9
6	18.85	23½	73.83	57	179.1	92	289.0
6½	20.42	24	75.40	58	182.2	93	292.2
7	21.99	24½	76.97	59	185.4	94	295.3
7½	23.56	25	78.54	60	188.5	95	298.5
8	25.13	26	81.68	61	191.6	96	301.6
8½	26.70	27	84.82	62	194.8	97	304.7
9	28.27	28	87.96	63	197.9	98	307.9
9½	29.84	29	91.11	64	201.1	99	311.0

Mathematical Tables

Tables 4-11, 4-12, and 4-13 are for convenient reference and will be found useful in numerous calculations.

Table 4-13. Logarithms of Numbers

No.	0	1	2	3	4	5	6	7	8	9	Diff.
10	00000	00432	00860	01284	01703	02119	02531	02938	03342	03743	415
11	04139	04532	04922	05308	05690	06070	06446	06819	07188	07555	379
12	07918	08279	08636	08991	09342	09691	10037	10380	10721	11059	344
13	11394	11727	12057	12385	12710	13033	13354	13672	13988	14301	323
14	14613	14922	15229	15534	15836	16137	16435	16732	17026	17319	298
15	17609	17898	18184	18469	18752	19033	19312	19590	19866	20140	281
16	20412	20683	20952	21219	21484	21748	22011	22272	22531	22789	264
17	23045	23300	23553	23805	24055	24304	24551	24797	25042	25285	249
18	25527	25768	26007	26245	26482	26717	26951	27184	27416	27646	234
19	27875	28103	28330	28556	28780	29003	29226	29447	29667	29885	222
20	30103	30320	30535	30750	30963	31175	31387	31597	31806	32015	212
21	32222	32428	32634	32838	33041	33244	33445	33646	33846	34044	202
22	34242	34439	34635	34830	35025	35218	35411	35603	35793	35984	193
23	36173	36361	36549	36736	36922	37107	37291	37475	37658	37840	185
24	38021	38202	38382	38561	38739	38917	39094	39270	39445	39620	177
25	39794	39967	40140	40312	40483	40654	40824	40993	41162	41330	170
26	41497	41664	41830	41996	42160	42325	42488	42651	42813	42975	164
27	43136	43297	43457	43616	43775	43933	44091	44248	44404	44560	158
28	44716	44871	45025	45179	45332	45484	45637	45788	45939	46090	153
29	46240	46389	46538	46687	46835	46982	47129	47276	47422	47567	148
30	47712	47857	48001	48144	48287	48430	48572	48714	48855	48996	143
31	49136	49276	49415	49554	49693	49831	49969	50160	50243	50379	138
32	50515	50651	50786	50920	51055	51189	51322	51455	51587	51720	134
33	51851	51983	52114	52244	52375	52504	52634	52763	52892	53020	130
34	53148	53275	53403	53529	53656	53782	53908	54033	54158	54283	126
35	54407	54531	54654	54777	54900	55023	55145	55267	55388	55509	122
36	55630	55751	55871	55991	56110	56229	56348	56467	56585	56703	119
37	56820	56937	57054	57171	57287	57403	57519	57634	57749	57864	116
38	57978	58093	58206	58320	58433	58546	58659	58771	58883	58995	113
39	59106	59218	59329	59439	59550	59660	59770	59879	59988	60097	110
40	60206	60314	60423	60531	60638	60746	60853	60959	61066	61172	107
41	61278	61384	61490	61595	61700	61805	61909	62014	62118	62221	104
42	62325	62428	62531	62634	62737	62839	62941	63043	63144	63246	102
43	63347	63448	63548	63649	63749	63849	63949	64048	64147	64246	99
44	64345	64444	64542	64640	64738	64836	64933	65031	65128	65225	98
45	65321	65418	65514	65610	65706	65801	65896	65992	66087	66181	96
46	66276	66370	66464	66558	66652	66745	66839	66932	67025	67117	95
47	67210	67302	67394	67486	67578	67669	67761	67852	67943	68034	92
48	68124	68215	68305	68395	68485	68574	68664	68753	68842	68931	90
49	69020	69108	69197	69285	69373	69461	69548	69636	69723	69810	88
50	69897	69984	70070	70157	70243	70329	70415	70501	70586	70672	86
51	70757	70842	70927	71012	71096	71181	71265	71349	71433	71517	84
52	71600	71684	71767	71852	71933	72016	72099	72181	72263	72346	82
53	72428	72509	72591	72673	72754	72835	72916	72997	73078	73159	81
54	73239	73320	73400	73480	73560	73640	73719	73799	73878	73957	80

Table 4-13. Logarithms of Numbers (Cont'd)

No.	0	1	2	3	4	5	6	7	8	9	Diff.
55	74036	74115	74194	74273	74351	74429	74507	74586	74663	74741	78
56	74819	74896	74974	75051	75128	75205	75282	75358	75435	75511	77
57	75587	75664	75740	75815	75891	75967	76042	76118	76193	76268	75
58	76343	76418	76492	76567	76641	76716	76790	76864	76938	77012	74
59	77085	77159	77232	77305	77379	77452	77525	77597	77670	77743	73
60	77815	77887	77960	78032	78104	78176	78247	78319	78390	78462	72
61	78533	78604	78675	78746	78817	78888	78958	79029	79099	79169	71
62	79239	79309	79379	79449	79518	79588	79657	79727	79796	79865	70
63	79934	80003	80072	80140	80209	80277	80346	80414	80482	80550	69
64	80618	80686	80754	80821	80889	80956	81023	81090	81158	81224	68
65	81291	81358	81425	81491	81558	81624	81690	81757	81823	81889	67
66	81954	82020	82086	82151	82217	82282	82347	82413	82478	82543	66
67	82607	82672	82737	82802	82866	82930	82995	83059	83123	83187	64
68	83251	83315	83378	83442	83506	83569	83632	83696	83759	83822	63
69	83885	83948	84011	84073	84136	84198	84261	84323	84386	84448	63
70	84510	84572	84634	84696	84757	84819	84880	84942	85003	85065	62
71	85126	85187	85248	85309	85370	85431	85491	85552	85612	85673	61
72	85733	85794	85854	85914	85974	86034	86094	86153	86213	86273	60
73	86332	86392	86451	86510	86570	86629	86688	86747	86806	86864	59
74	86923	86982	87040	87099	87157	87216	87274	87332	87390	87448	58
75	87506	87564	87622	87680	87737	87795	87852	87910	87967	88024	57
76	88081	88138	88196	88252	88309	88366	88423	88480	88536	88593	57
77	88649	88705	88762	88818	88874	88930	88986	89042	89098	89154	56
78	89209	89265	89321	89376	89432	89487	89542	89597	89653	89708	55
79	89763	89818	89873	89927	89982	90037	90091	90146	90200	90255	54
80	90309	90363	90417	90472	90526	90580	90634	90687	90741	90795	54
81	90849	90902	90956	91009	91062	91116	91169	91222	91275	91328	53
82	91381	91434	91487	91540	91593	91645	91698	91751	91803	91855	53
83	91908	91960	92012	92065	92117	92169	92221	92273	92324	92376	52
84	92428	92480	92531	92583	92634	92686	92737	92788	92840	92891	51
85	92942	92993	93044	93095	93146	93197	93247	93298	93349	93399	51
86	93450	93500	93551	93601	93651	93702	93752	93802	93852	93902	50
87	93952	94002	94052	94101	94151	94201	94250	94300	94349	94399	49
88	94448	94498	94547	94596	94645	94694	94743	94792	94841	94890	49
89	94939	94988	95036	95085	95134	95182	95231	95279	95328	95376	48
90	95424	95472	95521	95569	95617	95665	95713	95761	95809	95856	48
91	95904	95952	95999	96047	96095	96142	96190	96237	96284	96332	48
92	96379	96426	96473	96520	96567	96614	96661	96708	96755	96802	47
93	96848	96895	96942	96988	97035	97081	97128	97174	97220	97267	47
94	97313	97359	97405	97451	97497	97543	97589	97635	97681	97727	46
95	97772	97818	97864	97909	97955	98000	98046	98091	98137	98182	46
96	98227	98272	98318	98363	98408	98453	98498	98543	98588	98632	45
97	98677	98722	98767	98811	98856	98900	98945	98989	99034	99078	45
98	99123	99167	99211	99255	99300	99344	99388	99432	99476	99520	44
99	99564	99607	99651	99695	99739	99782	99826	99870	99913	99957	44

CHAPTER 5

Automatic Fire Protection Systems

The installation and maintenance of automatic fire protection systems is a particular branch of the pipe-fitting industry. Workmen who specialize in this field are called sprinkler fitters.

Sprinkler fitters serve an apprenticeship during which time they are taught to use tools and machinery common to the plumbing and pipe-fitting trades. They also learn the various codes and standards which apply to fire protection installations. Hundreds of lives and millions of dollars are saved each year because fire sprinkler systems are installed in hotels, motels, commercial and public buildings. Automatic fire sprinkler systems quench small fires before they can become roaring infernos. Some typical applications of fire protection systems (sprinkler, foam, etc.) are hotels, motels, apartments, department stores, furniture stores, grocery stores, schools, public buildings, warehouses, hi-rise buildings, flammable liquid loading facilities, aircraft hangars, and off-shore oil platforms.

The scope of work involved includes, but is not limited to, installation of private fire service (water) mains, installation of piping, pipe hangers and supports, valves, sprinkler heads, foam systems and generators, pumps, monitor systems, air compressors, tanks, and also regular inspections, testing, and servicing of equipment.

Sprinkler Systems

A sprinkler system designed for fire protection is an integrated system of underground and overhead piping which includes one or more automatic water supplies. The portion of the sprinkler system above ground is a network of specially sized or hydraulically designed piping installed in a building, structure, or area, generally overhead, and to which sprinklers are attached in a systematic pattern. The valve controlling each system riser is located in the system riser or its supply piping. Each sprinkler riser includes a device for actuating an alarm when the system is in operation.

The system is usually activated by heat from a fire and discharges water over the fire area. Sprinkler systems can be classified into two main types:

1. Wet pipe systems.
2. Dry pipe systems.

Further variations of these systems are:

3. Pre-action systems.
4. Deluge systems.
5. Combined dry pipe and pre-action systems.

Wet Pipe System

Description

A wet pipe sprinkler system is fixed fire protection using piping filled with pressurized water and activated by fusible sprinklers for the control of fire.

Application

A wet pipe sprinkler system may be installed in any structure not subject to freezing temperatures to automatically protect the structure, contents, and/or personnel from loss due to fire. The structure must be substantial enough to support the piping system filled with water.

Operation of a Wet Pipe System

When a fire occurs, the heat produced will fuse a sprinkler, causing water to flow. The alarm valve clapper is opened by the flow and allows pressurized water to fill the retarding chamber. The flow overcomes the retarding chamber's small capacity drain and fills the alarm line. This, in turn, closes the pressure switch, sounds an electric alarm, and operates the mechanical water motor alarm. If a water-flow indicator is used in the system piping, it also is activated by the water flow.

The paddle which normally lies motionless inside the pipe is forced up, thereby activating the pneumatic time delay mechanism which closes or opens a micro-switch after the preset retard time has elapsed. This action causes an electric alarm to sound. All alarms will continue to sound as long as there is a flow of water in the system. The water will flow until it is shut off manually. Components of a wet pipe system are shown in Fig. 5-1.

Dry Pipe System

Description

A dry pipe sprinkler system is a fire protection system which utilizes water as an extinguishing agent. The system piping from the dry pipe valve to the fusible sprinklers is filled with pressurized air or nitrogen.

An air check system is a small dry system which is directly connected to a wet pipe system. The air check system uses a dry valve and an air supply but does not have a separate alarm. The alarm is provided by the main alarm valve.

Application

A dry pipe system is primarily used to protect unheated structures or areas where the system is subject to freezing temperatures. Under such circumstances it may be installed in any structure to automatically protect the structure contents and/or personnel from loss due to fire. The structure must be substantial enough to support the system piping when filled with water.

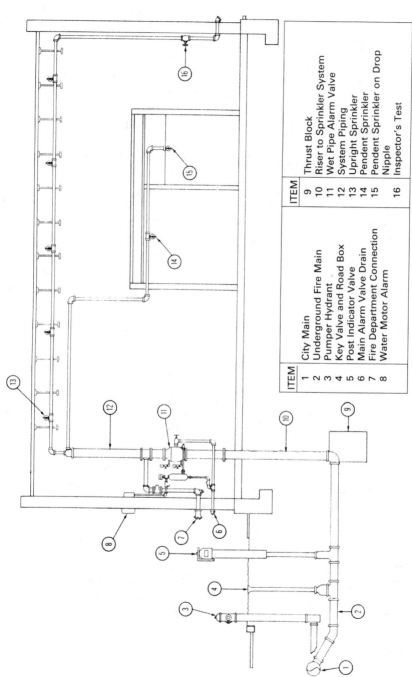

ITEM		ITEM	
1	City Main	9	Thrust Block
2	Underground Fire Main	10	Riser to Sprinkler System
3	Pumper Hydrant	11	Wet Pipe Alarm Valve
4	Key Valve and Road Box	12	System Piping
5	Post Indicator Valve	13	Upright Sprinkler
6	Main Alarm Valve Drain	14	Pendent Sprinkler
7	Fire Department Connection	15	Pendent Sprinkler on Drop
8	Water Motor Alarm	16	Nipple
		16	Inspector's Test

Fig. 5-1.   Components of a wet pipe sprinkler system. *(Courtesy The Viking Co.)*

Operation of a Dry Pipe System

The components of a dry pipe sprinkler system are shown in Fig. 5-2. Note that as explained in the operation of a wet pipe system, the components of a dry pipe system may vary due to the application of different sets of standards. The system shown in Fig. 5-2 is only one possible arrangement of a dry pipe system. Additional components of a dry pipe system are an adequate water supply taken from a city main, elevated storage tank, ground storage reservoir and fire pump, deep well pump, and storage tank.

Underground System

1. Piping—Cast iron, ductile iron, or cement asbestos.
2. Control Valves—Post indicator valves (PIV).
3. Valve Pit—Usually required when multiple sprinkler systems are serviced from a common underground system taking supply from a city main: (2) OS&Y (open stem and yoke) valves, check valve or detector check, fire department connection ($2\frac{1}{2}'' \times 2\frac{1}{2}'' \times 4''$ hose connection, 4 in. check valve with ball drip). Check local building codes for equipment and building requirements. A backflow preventer, full flow meter, or combinations of equipment may be locally required.
4. Auxiliary Equipment—Fire hydrants with two $2\frac{1}{2}'' \times 2\frac{1}{2}''$ outlets for hose line use and one 4 in. outlet (for connection to fire department pump truck) and hose and equipment houses. Check with the fire department servicing the location for hydrant and hose thread requirements.

Deluge Systems

Another type of sprinkler system is the deluge system. A deluge system is an empty pipe system which is used in high-hazard areas or in areas where fire may spread rapidly. It can also be used to cool surfaces such as tanks, process lines, or transformers. In this type of application, open sprinklers or spray nozzles are employed for water distribution. The deluge valve is activated by a release system employing manual, fixed temperature, rate of temperature

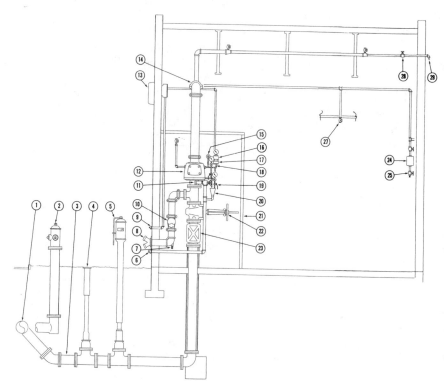

ITEM	DESCRIPTION	ITEM	DESCRIPTION	ITEM	DESCRIPTION
1	City Main	12	Dry Pipe Valve	23	Check Valve
2	Pumper Type Fire		Water Motor Alarm	24	Drum Drip
	Hydrant	13	Cross Main	25	Drain Valve & Plug
3	Underground Fire	14	Air Press. Main.	26	Upright Sprinkler
	Main	15	Device	27	Pendent Sprinkler
4	Key Valve & Road		Accelerator	28	Inspector's Test
	Box	16	(optional)	29	Valve
5	Post Indicator Valve	17	Pressure Switch		Inspector's Test
6	Main Drain	18	(hidden)		Drain
7	Ball Drip		Alarm Line Strainer		
8	Fire Dept.	19	(hidden)		
9	Connection		Alarm Test Valve		
	Water Motor Alarm	20	Drain Cup		
10	Drain	21	Dry Pipe Valve		
11	Check Valve	22	House		
	Main Drain Valve		O.S. & Y. Valve		
			(optional)		

Fig. 5-2. Components of a dry pipe sprinkler system. *(Courtesy The Viking Co.)*

rise, radiation, smoke or combustion gases, hazardous vapors, or pressure increase. Once a deluge system is tripped, water or other extinguishing agent flows through all spray nozzles and/or sprinkler heads simultaneously. Two applications of deluge systems are flammable liquid loading facilities and aircraft hangar fire protection.

Flammable Liquid Loading Facilities

Truck and rail loading facilities for flammable or combustible liquids routinely handle large volumes of potentially dangerous material in complete safety. Vehicles, however, may collide or hit the loading structure, rupturing tanks or pipes. Hoses, nozzles, and valves that are in constant use may break or malfunction, and operator error is always a hazard. Any of these failures could result in a spill which, if ignited, would cause a disastrous fire.

Since large volumes of fuel are present, it is likely that a fire would develop rapidly and endanger personnel, vehicles, cargoes, and the unprotected steel loading structure. The loading site is almost always curbed or otherwise contained since it is not usually permissible to wash the fire to another area. For that reason the fire must be quickly extinguished, and not allowed to re-flash.

Control Strategy

A widely used and economical way to handle the hazard is a foam-water deluge sprinkler system. The structure and the upper portion of the vehicles are kept cool by a water spray, but since water only is not effective in the extinguishment of buoyant, burning liquids, a foaming agent must be introduced into the water.

Because a fire may be expected to develop very quickly, combination rate-of-rise and fixed temperature detectors are normally employed. Backup manual activation should also be provided.

Pipelines carrying flammable liquid or fuel to the loading area may be equipped with shutoff valves which are automatically closed when the fire detectors operate. Pumps associated with this system may also be interlocked to shut off. A foam deluge system is shown in Fig. 5-3.

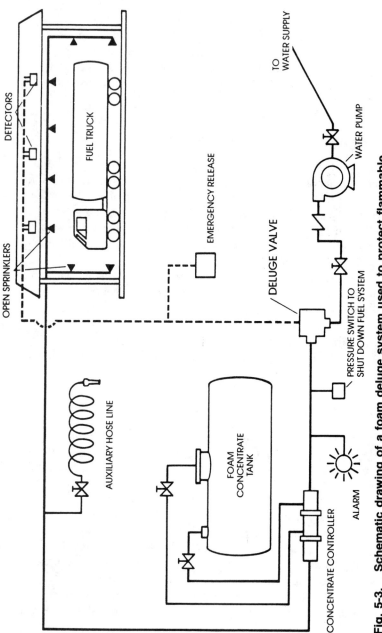

Fig. 5-3. Schematic drawing of a foam deluge system used to protect flammable liquid loading facilities. *(Courtesy The Viking Co.)*

Foam-Water Deluge System

Automatic detection is accomplished by activating either fixed-temperature or rate-of-temperature-rise mechanisms. Manual emergency releases are also provided. When the detection system operates, the deluge valve opens and supplies water to the sprinkler piping. Aqueous film-forming foam (AFFF) concentrate is introduced into the water by the proportioner and is mixed in the sprinkler piping. Unsealed, upright sprinklers provide for the structure, the top of the vehicles, and the surrounding area. Unsealed sprinklers located on the lower structure deliver foam to the underside of the vehicle and over the ground below. Hand-held hose lines controlled by manual valves are also fed from the sprinkler piping. Water flow alarms sound when the sprinkler piping is flooded. Flammable liquid supply may be stopped by pump shut-down or an automatic shut-off valve actuated by a pressure switch in the sprinkler piping.

The components of a fire protection system for flammable liquid loading facilities are shown in Fig. 5-3.

Other Equipment and Options

Protein foam may be used in place of AFFF. If protein foam is used, special foam water sprinklers must replace standard open sprinklers. If large quantities of water and foam concentrate are required, pumps, which are started by a signal from the detection system, are usually employed for both liquids. Flow control valves with downstream pressure regulators may be used in place of deluge valves to conserve water and foam concentrate.

In the event that release is in a freezing area, a pneumatic release system must be used and a suitable air supply provided. Electrical fixed-temperature and rate-of-temperature-rise detection may be used in place of mechanical detection, but such detectors may be required to meet hazardous area standards.

Infrared or ultraviolet detection may also be used, but consideration must be given to the increased potential for false trips.

Aircraft Hangar Fire Protection

Today's huge aircraft hangar represents a large and concentrated loss potential. The primary hazard is the ignition of a fuel spill involving an aircraft located within the hangar. An aircraft is extremely vulnerable to damage caused by heat sources producing skin temperatures above 400°F (250°C). The hangar itself is susceptible to fire damage, since usual construction includes an unprotected steel-supported roof which will buckle and possibly collapse on exposure to high temperatures.

Control Strategy

Rapid control and extinguishment of fire is vital if loss is to be minimized. Even when it is not possible to save an aircraft, it is necessary to control and extinguish the fire to protect the hangar. Since buoyant, flammable liquid is the principal fire hazard, extinguishing systems employing a foam-water agent are the preferred means of protection. The aircraft and the building structure are cooled by the water while the burning fuel is smothered by the foam. Water alone generally means higher density requirements, resulting in higher costs and lower efficiency.

It is not practical to mount fixed fire protection directly under the aircraft where fire is most likely to occur. Therefore, three types of extinguishing systems are normally provided for complete coverage, all employing foam. First, a monitor nozzle system, automatically activated by high-sensitivity optical detectors, is used to protect the underside of the aircraft, which is shielded from standard foam-water sprinkler protection installed at the hangar roof. Second, the roof sprinkler system, activated by rate-of-rise detectors, provides general area protection, usually deluge. This foam-water system provides building protection in the event the monitor nozzle system fails to control a fire. Third, hand-held hose lines with foam nozzles are provided for manual fire-fighting operations.

Because building protection is provided by a deluge system, large amounts of water and foam concentrate are required. These must be properly managed to avoid exhausting the supply. Pumping facilities are almost always required to meet high volume demands.

Operation of a Foam-Water Deluge System

Initial detection is normally accomplished by the monitor detection system employing either infrared or ultraviolet detectors. These systems are capable of detecting a fire in its early stages. However, false trips with such detection systems are a possibility and must be seriously considered. Manual activation of the monitor system is also provided.

In a typical system actuation, a signal from the monitor detector is received by the release control panel (which may be interlocked to the water-and-foam concentrate supply pump starters). This signal opens both flow control valves supplying water and foam to the monitor nozzles and sounds an alarm. Foam concentrate is proportioned into the water and mixed in the supply piping. The flow control valve operates both as a deluge valve and a downstream pressure-regulating valve to insure the correct proportion of water and foam concentrate regardless of varying supply pressure. The monitor nozzles move in a programmed path to cover the aircraft, the under-aircraft areas, and areas that might contain a fuel spill. Hose lines supplied by individual manual valves are located at convenient places for manual fire fighting.

In the event the monitor nozzles do not control the fire, the temperature will increase at the roof. This condition will trip a release, which is activated by either fixed-temperature or rate-of-temperature-rise conditions, causing the flow control valve to open. Unsealed, upright sprinklers then distribute the foam over the general area. Operation of the roof system will trip the monitor system in the event that it has not already functioned. That operation may also be designed to actuate systems in adjacent areas.

A schematic drawing of an aircraft hangar fire protection system is shown in Fig. 5-4.

Other Equipment and Options

Protein or fluoroprotein foam may be used in place of AFFF. If such foam is used, special foam-water sprinklers must also be used in place of standard open sprinklers. In the event that the release is in a freezing area, pneumatic operation must be used and a suitable air supply provided.

Electrical fixed-temperature and rate-of-temperature-rise detection utilizing a release control panel may be used in place of mechanical detection at the hangar roof.

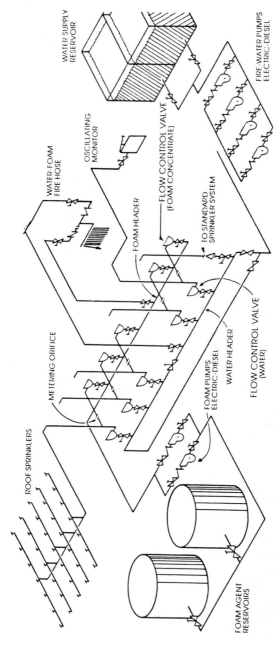

Fig. 5-4.   Schematic drawing of an aircraft hangar fire protection system. *(Courtesy The Viking Co.)*

Materials Used in Sprinkler Fitting

UNDERGROUND PIPING

Cast iron.
Ductile iron.
Cement asbestos.

ABOVE-GROUND PIPING

Standard practice in most areas is to use Schedule 40 steel pipe and threaded cast-iron fittings for mains and branch piping. 2½ in. I.D. (inside diameter) and larger valves and fittings are usually flanged and joined to threaded pipe by companion flanges.

Other methods of joining pipe and fittings are: welding, brazing, and mechanically grooved fittings used with cut- or rolled-on grooved pipe. Generally, the installation costs of these methods are higher than entailed by the use of threaded and/or flanged pipe and fittings.

Post indicator valves (PIV) indicate by visible words whether the main to the sprinkler system is open or shut. An OS&Y valve indicates by the visible position of the stem whether the valve is open or shut. Three types of sprinkler heads are shown in Fig. 5-5.

Fig. 5-5. Three types of sprinkler heads. *(Courtesy The Viking Co.)*

Fig. 5-6.   A double-check backflow preventer. *(Courtesy The Viking Co.)*

Backflow Preventers

When fire protection systems are connected to potable water supplies, check valves should be installed in the system piping. This is to prevent backflow of potentially polluted water into the potable water supply. A double-check (valves) backflow preventer is shown in Fig. 5-6.

Fire department pumper trucks arrive at a fire and if necessary connect a suction hose to a nearby fire hydrant. A pressure hose from the truck's pump can, if necessary, be connected to the fire department connection (Fig. 5-2, Item 8). Then, if water at higher pressure is needed, the pumper truck can pump water into the fire protection system piping. The pumper truck is capable of building pressure in excess of water main pressure and, if the backflow preventer were not installed in the fire protection system piping, could force potentially polluted water into the potable water main.

CHAPTER 6

Physics for Plumbers and Pipe Fitters

By definition, physics is *the science or group of sciences that treats of the phenomena associated with matter in general, especially in its relations to energy, and of the laws governing these phenomena, excluding the special laws and phenomena peculiar to living matter (biology) or to special kinds of matter (chemistry).*

Physics is generally considered to include the study of:

1. The constitution and properties of matter.
2. Mechanics.
3. Acoustics.
4. Heat.
5. Optics.
6. Electricity and magnetism.

As sometimes used in a limited sense, physics embraces only the last four divisions; more generally, it includes all the physical sciences.

According to Barker, physics regards matter solely as the vehicle of energy. From this point of view, physics may be defined as *that department of science whose province it is to investigate all those phenomena of nature that depend either upon the transference of energy from one portion of matter to another, or upon its transformation into any of the forms it is capable of assuming.* In a word, physics may be regarded as the science of energy, precisely as chemistry may be regarded as the science of matter.

The scope of physics extends considerably beyond what is important to the plumber. Only relevant subjects will be presented here. In this connection, the plumber should thoroughly study this chapter, and should understand not only such things as why pipes burst in freezing weather or why water circulates in hot-water heating systems, but also should know the reasons for all the various phenomena commonly observed at work. For instance, he should know why pipes become air bound; why air chambers on pumps fill with water; why a boiler water gauge does not register the true water level; why a bucket-valve pump delivers more than its displacement, etc.

Measurements

According to Plato, physics begins with measurements. In fact, if arithmetic, mensuration, and weighing are taken away from any art, that which remains will not be much.

There are three fundamental kinds of measurements:

1. Length.
2. Mass.
3. Time.

In addition to these, there are *derived* measurements of:

1. Area.
2. Volume.

These are called *derived* because they are the products of two and three lengths. Various units are used for these measurements. The plumber uses the ordinary unit such as inches, pounds, and seconds for fundamental measurements, and uses square inches and cubic inches for derived measurements. In addition to measuring the size or weight of an object, other measurements are necessary in physics, such as the measurement of pressure and temperature. Such measurements are indicated by instruments provided with arbitrary scales divided into standard divisions, each division standing for a unit of pressure, temperature, etc.

Most commonly used for measurement are the wood folding

Fig. 6-1.   A six-foot folding rule.

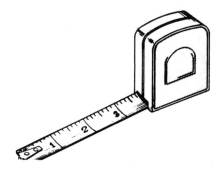

Fig. 6-2.   A steel tape.

rule (Fig. 6-1) and the steel tape. The steel tape (Fig. 6-2) is enclosed, as the name implies, in a metal case with the measuring tape folding into the device for ease of carrying and use. Both the folding and tape rule are usually 6 ft. long, although many of the steel tapes are available in 8- and 10-ft. lengths, making floor-to-ceiling measurements much easier.

Water

Water is a *compound of hydrogen and oxygen in the proportion of 2 parts by weight of hydrogen to 16 parts by weight of oxygen.* Since the atom of oxygen is believed to weigh 16 times as much as the atom of hydrogen, the molecule of water is said to contain 2 atoms of hydrogen and 1 atom of oxygen, being represented by the formula H_2O.

Under the influence of temperature and pressure, this substance (H_2O) may exist as:

1. A solid.
2. A liquid.
3. A gas.

As a solid, it is called *ice*; as a liquid, *water*; as a gas, *steam*. Water at its maximum density (39.1°F) will expand as heat is added, and it will also expand slightly as the temperature falls from this point, as illustrated in Fig. 6-3. Water will freeze at 32°F and boil at 212°F, when the barometer reads 29.921 in. of mercury.

The boiling point of water is not the same in all places. It decreases as the altitude increases; at an altitude of 5000 ft., water will boil at a temperature of 202°F. An increase of pressure will elevate the boiling point of water. At maximum density, the weight of a cu. ft. of water is generally taken as 62.425 lbs. One U.S. gallon (231 cu. in.) of water weighs $8\frac{1}{3}$ lbs. The figure $8\frac{1}{3}$ is correct when the water is at a temperature of 65°F. The pressure of water varies with the head, and is equal to .43302 lbs. per sq. in. for every foot of (static) head.

Heat

By definition, heat *is a form of energy known by its effects.* These effects are indicated through touch and feel as well as by the ex-

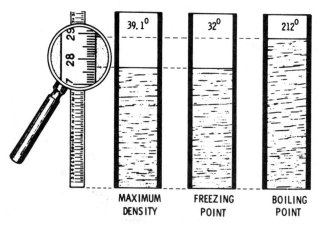

Fig. 6-3.   The effect on water of various temperatures.

pansion, fusion, combustion, or evaporation of the matter upon which it acts. Temperature is that which indicates how hot or cold a substance is; a measure of *sensible heat*.

Sensible heat is that heat which produces a rise in temperature as distinguished from latent heat. *Latent heat* is that quantity of heat required to change the *state* or condition under which a substance exists without changing its temperature. Thus, a definite quantity of heat must be transferred to ice at 32° to change it into water at the same temperature.

Specific heat is the ratio of the quantity of heat required to raise the temperature of a given weight of any substance one degree to the quantity of heat required to raise the temperature of the same weight of water from 62° to 63°F. When bodies of unequal temperatures are placed near each other, heat leaves the hot body and is absorbed by the colder body until the temperature of each is equal. This is called a transfer of heat.

The rate by which the heat is absorbed by the colder body is proportional to the difference of temperature between the two bodies. The greater the difference of temperature, the greater the rate of flow of the heat. The transfer of heat takes place by radiation, conduction, or convection. Thus, in a boiler, heat is given off from the furnace fire in rays which radiate in straight lines in all directions, being transferred to the crown and sides of the furnace by radiation; it passes through the plates by conduction, and is transferred to the water by convection (that is, by currents).

Bodies expand by the action of heat. For instance, boiler plates are riveted with red-hot rivets in an expanded state; on cooling, the rivets contract and draw the plates together with great force, making a tight joint. An exception to the rule, it should be noted, is water, which contracts as it is heated from the freezing point 32°F, to the point of maximum density at 39.1°; at other temperatures it expands.

Heat and Work

Heat develops *mechanical force* and *motion*; hence, it is *convertible into mechanical work*. Heat is measured by a standard unit called the British unit of heat. The *British thermal unit* is equal to $\frac{1}{180}$ of the heat required to raise the temperature of one pound of water from 32° to 212°F. It should be noted that this is the definition adapted in this work for the British thermal unit (Btu), corresponding to the unit used in the Marks and Davis steam tables, which is now the recognized standard.

Work

By definition, work is *the overcoming of resistance through a certain distance by the expenditure of energy.* Work is measured by a standard unit called the *foot-pound.* A foot-pound is *the amount of work done in raising 1 lb. 1 ft.*, or in overcoming a pressure of 1 lb. through a distance of 1 ft. Thus, if a 5-lb. weight is raised 10 ft., the work done is 5 × 10 = 50 foot-pounds.

Joule's Experiment

It was shown by experiments made by Joule in 1843–50 that 1 *unit of heat = 772 units of work.* This is known as the *mechanical equivalent of heat,* or Joule's equivalent.

Experiments by Professor Rowland (1880) and others give higher figures; 778 is generally accepted, but 777.5 is probably more nearly correct, the value 777.52 being used by Marks and Davis in their steam tables. The value 778 is sufficiently accurate for ordinary calculations.

Energy

By definition, *energy is stored work*; that is, the ability to do work, or in other words, to move against resistance. A body may possess energy whether it does any work or not, but no work is ever done except by the expenditure of energy. There are two kinds of energy:

1. Potential.
2. Kinetic.

Potential energy is energy due to position, as represented, for instance, by a body of water stored in an elevated reservoir, and capable of doing work by means of a water wheel.

Kinetic energy is energy due to momentum; that is, the energy of a moving body.

Conservation of Energy

The doctrine of physics is that energy can be transmitted from one body to another or transformed in its manifestations, but *may neither be created nor destroyed.*

Power

By definition, power is the *rate* at which work is done; in other words, it is work divided by the *time* in which it is done. The unit of power in general use is the *horsepower*, which is defined as 33,000 foot-pounds per minute. One horsepower is required to raise a weight of:

33,000 pounds	1 foot in one minute
3300 pounds	10 feet in one minute
330 pounds	100 feet in one minute
33 pounds	1000 feet in one minute
3.3 pounds	10,000 feet in one minute
1 pound	33,000 feet in one minute, etc.

Pressure

By definition, pressure is *a force, in the nature of a thrust, distributed over a surface*; in other words, the kind of force with which a body tends to expand or resist an effort to compress it. Pressure is usually stated in pounds per square inch, meaning that a pressure of a given number of pounds is distributed over each square inch of surface. This should be very clearly understood as further explained in Fig. 6-4.

Atmospheric pressure is the force exerted by the weight of the atmosphere on every point with which it is in contact. At sea level, this pressure is taken at 14.7 lbs. per sq. in. for ordinary calculations. We do not feel the atmospheric pressure because air presses the body both externally and internally so that the pressures in different directions balance. Atmospheric pressure varies with the elevation. The pressure decreases approximately one-half pound for every 1000 ft. of ascent. It is measured by an instrument called the barometer.

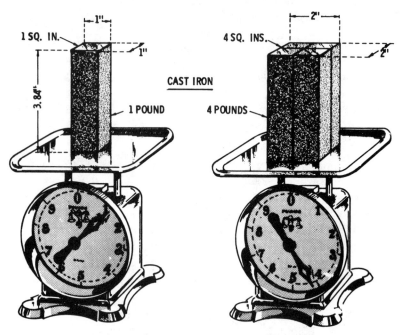

Fig. 6-4. Pressure per square inch.

Barometer

By definition, a barometer is an instrument for measuring the pressure of the atmosphere, as shown in Fig. 6-5. The instrument consists of a glass tube 33 to 34 in. high, sealed at the top, filled with pure mercury, and inverted in an open cup of mercury. A graduated scale on the instrument permits observations of the fluctuations in the height of the mercury column. It is highest when the atmosphere is dry, weighing more then than when saturated with aqueous vapor, which is lighter than air. The height of barometric measurement is about 30 in.

The column of mercury remains suspended at this height because the weight of a column of mercury 30 in. high is the same as the weight of a like column of air about 50 miles high.

Pressure Scales

The term *vacuum*, strictly speaking, is defined as a *space devoid of matter*. This is equivalent to saying *a space in which the pressure*

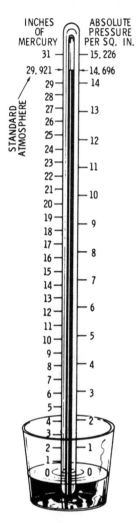

Fig. 6-5. **A barometer illustrating the relationship between inches of mercury and absolute pressure in lbs. per sq. in.**

is zero. According to common usage, it means *any space in which the pressure is less than that of the atmosphere.*

This gives rise to two scales of pressure:

1. Gauge.
2. Absolute.

When the hand of a steam gauge is at zero, the pressure actually existing is 14.74 lbs. (referred to a 30-in. barometer), or that of the atmosphere. The scale in the gauge is not marked at this point (14.74 lbs.), but at zero because, in the steam boiler as well as any other vessel under pressure, the important measurement is the difference of pressure between the inside and outside. This difference of pressure, or the effective pressure for doing work, is called the *gauge pressure* because it is measured by the gauge on the boiler.

The second pressure scale is known as *absolute pressure* because it gives the actual *pressure above zero*. In all calculations relative to the expansion of steam, the absolute pressure scale must be used. Gauge pressure is expressed as absolute pressure by adding 14.74, or for ordinary calculations, 14.7 lbs. Thus, 80 lbs. gauge pressure = 80 + 14.74 = 94.74 lbs. absolute pressure. Absolute pressure is expressed as gauge pressure by subtracting 14.7. Thus 90 lbs. absolute pressure = 90 − 14.7 = 75.3 lbs. gauge pressure.

The pressures below atmospheric pressure are usually expressed in lbs. per sq. in. when making calculations, or "inches of mercury" in practice. Thus, in the engine room, the expression "28 in. of vacuum" would signify an absolute pressure in the condenser of .946 lb. per sq. in. absolute. In other words, the mercury in a mercury column connected to a condenser having a 28-in. vacuum would rise to a height of 28 in., representing the difference between the pressure of the atmosphere and the pressure in the condenser of 14.73 − .946 = 13.784 lbs. referred to a 30-in. barometer.

Pressure in lbs. per sq. in. is obtained by multiplying the barometer reading by .49116. Thus, a 30-in. barometer reading signifies a pressure of .49116 × 30 = 14.74 lbs. per sq. in.

Table 6-1 gives the pressure of the atmosphere in pounds per square inch for various readings of the barometer.

Table 6-1. Atmospheric Pressure per Square Inch

Barometer (ins. of mercury)	Pressure (lbs. per sq. in.)	Barometer (ins. of mercury)	Pressure (lbs. per sq. in.)
28.00	13.75	29.291	14.696
28.25	13.88	30.00	14.74
28.50	14.00	30.25	14.86
28.75	14.12	30.50	14.98
29.00	14.24	30.75	15.10
29.25	14.37	31.00	15.23
29.50	14.49		
29.75	14.61		

Rule. Barometer in inches of mercury × .49116 = lbs. per sq. in.

Table 6-1 is based on the standard atmosphere which, by definition, equals 29.921 in. of mercury, which equals 14.696 lbs. per sq. in., or 1 in. of mercury = 14.696 ÷ 29.921 = .49116 lbs. per sq. in.

Thermometers

This term is generally applied to a glass tube terminating in a bulb charged with a liquid, usually mercury or colored alcohol. The liquid contracts or expands with changes of temperature, falling or rising in the tube against which is placed a graduated scale. The common scale is Fahrenheit, on which zero is the temperature of a mixture of salt and snow, 32° that of melting ice, and 212° that of boiling water. The Celsius and Reaumur scales, from the temperature of melting ice to that of boiling water, have 100 graduations and 80 graduations respectively. The Celsius is usually called the Centigrade thermometer. The latent heat varies with the boiling point —it decreases as the pressure rises.

Steam

By definition, *steam is the hot invisible vapor given off by water at its boiling point.* The visible white vapor popularly known as steam is not steam but a collection of fine watery particles formed by the condensation of steam.

Steam is said to be:

1. *Saturated* when its temperature corresponds to its pressure.
2. *Superheated* when its temperature is above that due to its pressure.
3. *Gaseous steam* or *steam gas* when it is highly superheated.
4. *Dry* when it contains no moisture. It may be either saturated or superheated.
5. *Wet* when it contains intermingled mist or spray, its temperature corresponding to its pressure.

Steam exists when there is the proper relation between the temperature of the water and the external pressure. For instance, for a given temperature of the water, there is a certain external pressure above which steam will not form. Steam is produced by heating water until it reaches the *boiling point*. The latent heat of steam is the amount of heat required to change one pound of water into steam at the same temperature.

Thus, if heat is applied to a pound of pure water having a temperature of 212°F, steam will be formed and in a short time all of the water will be evaporated. If the temperature of the steam so formed is taken, the thermometer will register the same as the boiling water, which is 212°. It has been accurately determined by experiment that 970.4° of heat, or heat units, must be applied to a pound of boiling water to change it into steam at the same temperature, and this heat is called the latent heat of steam.

The various states of steam can be seen in the operation of a safety valve by closely observing it when blowing off pressure. For instance, when the safety valve of a boiler furnishing superheated steam blows off, a very interesting phenomenon can be observed. Very close to the valve the escaping gas is entirely invisible, being superheated at this point. Farther away, the outline of the ascending column is seen, the interior being invisible and gradually becoming "foggy." As the vapor ascends, it gradually reduces in temperature, and the steam becomes saturated and supersaturated, or wet, reaching the white state a little farther away where it is popularly and erroneously known as "steam." Steam is invisible. The reason the so-called wet steam can be seen is that wet steam is a mechanical mixture made up of saturated steam (which is invisible), which holds in suspension a multiplicity of fine water globules formed by condensation. It is the collection of water globules or condensate that is visible.

Mechanical Powers

By definition, the mechanical powers are *mechanical contrivances that enter into the composition or formation of all machines*. They are:

1. The lever.
2. The wheel and axle.
3. The pulley.
4. The inclined plane.

5. The screw.
6. The wedge.

These can, in turn, be reduced to three classes:

1. A solid body turned on an axis.
2. A flexible cord.
3. A hard and smooth inclined surface.

The mechanism of the wheel and axle, and of the pulley, merely combines the principle of *the lever* with the tension of the cords. The properties of the screw depend entirely upon those of the lever and the inclined plane; and the case of the wedge is analogous to that of a body sustained between two inclined planes. They all depend for their action upon what is known as the *principle of work*, one of the important principles in mechanics and in the study of machine elements.

The principle of work states that, neglecting frictional or other losses, *the applied force multiplied by the distance through which it moves equals the resistance overcome multiplied by the distance through which it is overcome.* A force acting through a given distance can be made to overcome a greater force acting as a resistance through a shorter distance. No possible arrangement can be made to overcome a greater force through the same distance. The principle of work may also be stated as follows:

Work put into a machine = lost work + work done
by the machine

The principle holds true in every case. It applies equally to a simple lever, the most complex mechanism, or to a so-called *perpetual motion* machine. No machine can be made to perform work unless a somewhat greater amount (enough to make up for the losses) is applied by some external agent. In a perpetual motion machine, no such outside force is supposed to be applied; it is therefore against the laws of mechanics.

The Lever

The lever consists of an inflexible bar or rod, which is supported at some point, and is freely movable about that point as a center of motion. In the lever, three points are to be considered; the *fulcrum*

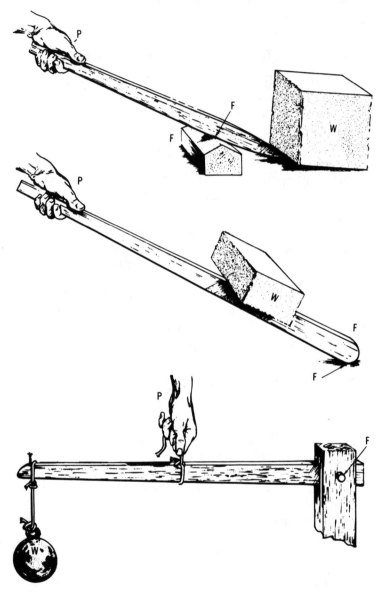

Fig. 6-6.   Three kinds of levers.

or point about which the lever turns, the point where the *force* is applied, and the point where the *weight* is applied. There are three varieties of the lever, as shown in Fig. 6-6. They differ according to where the *fulcrum*, the *weight*, or the *power* is respectively placed between the other two, but the action in every case is reducible to the same principle and the same general rule applies to them all. The following general rule holds for all classes of levers:

> **Rule.** The force *P*, multiplied by its distance from the *fulcrum*, is equal to the load *W* multiplied by its distance from the fulcrum. That is:

$$\text{Force} \times \text{distance} = \text{load} \times \text{distance}$$

Example—What force applied at 3 ft. from the fulcrum will balance a weight of 112 lbs. applied at a distance of 6 in. from the fulcrum?

Here, the distances or leverages are 3 ft. and 6 in. The distance must be of the same denomination; hence, reducing ft. to in., $3 \times 12 = 36$ in.

Applying the rule:

$$\text{Force} \times 36 = 112 \times 6$$

$$\text{Force} = \frac{112 \times 6}{36} = 18.67 \text{ or } 18\tfrac{2}{3} \text{ lbs.}$$

This solution holds for all levers illustrated in Fig. 6-7.

The Pulley

In its simplest form a pulley consists of a grooved wheel, called a sheave, turning within a frame by means of a cord or rope. It works in contact with the groove in order to transmit the force applied to the rope in another direction, as shown in Fig. 6-8. Pulleys are divided into *fixed* and *movable*. In the fixed pulley, no mechanical advantage is gained, but its use is of the greatest importance in accomplishing the work appropriate to the pulley, such as raising water from a well. The *movable* pulley, by distributing the weights into separate parts, is attended by mechanical advantages proportional to the number of points of support.

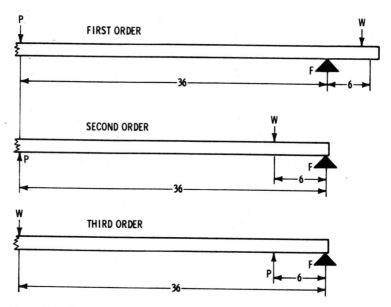

Fig. 6-7.   The three orders of levers.

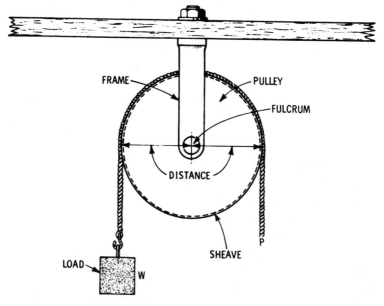

Fig. 6-8.   A simple pulley.

Combinations of pulleys are arranged with several sheaves in one frame to form a *block* to increase the load that may be lifted per unit of force applied; in other words, to increase the leverage. All such arrangements are virtually equivalents of the lever. The following rule expresses the relation between the force and load.

> **Rule.** The load capable of being lifted by a combination of pulleys is equal to the force × the number of ropes supporting the lower or movable block.

The Inclined Plane

This mechanical power consists of an inclined flat surface upon which a weight may be raised, as shown in Fig. 6-9. By such substitution of a sloping path for a direct upward line of ascent, a given weight can be raised by another weight weighing less than the weight to be raised. The inclined plane becomes a mechanical power in consequence of its supporting part of the weight, and leaving only a part to be supported by the power. Thus, the power has to encounter only a portion of the force of gravity at a time, a portion which is more or less according to how much the plane is elevated. The following rule expresses these relations:

> **Rule.** As the applied force P is to the load W, so is the height H to the length of the plane W. That is:

$$\text{Force : load} = \text{height : plane length}$$

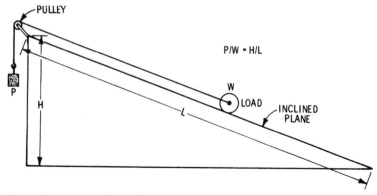

Fig. 6-9.   An inclined plane.

Example—What force (P) is necessary to raise a load of 10 lbs. if the height is 2 ft. and the plane is 12 ft.?

Substituting in the equation:

$$P : 10 = 2 : 12$$
$$P \times 12 = 2 \times 10$$
$$P = \frac{10 \times 2}{12} = \frac{20}{12} = 1\tfrac{2}{3} \text{ lbs.}$$

The Screw

This is simply an inclined plane wrapped around a cylinder. The evolution of a screw from an inclined plane is shown in Fig. 6-10. The distance between two consecutive coils, measured from center to center or from upper side to upper side (literally the height of the inclined plane for one revolution), is the *pitch* of the screw. The screw is generally employed when severe pressure is to be exerted through small distances. A screw in one revolution will descend a distance equal to its pitch, or the distance between two threads. The force applied to the screw will move through (in the same time) the circumference of a circle whose diameter is twice the length of the lever.

Rule. As the applied force is to the load, so is the pitch to the length of the thread per turn. That is:

Applied force : load = pitch : length of thread per turn

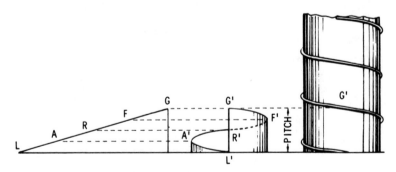

Fig. 6-10. Evolution of a screw.

Example—If the distance between the threads or pitch is ¼ in. and a force of 100 lbs. is applied at the circumference of the screw, what weight will be moved by the screw, if the length of the thread per turn of the screw is 10 in.?

Substituting in the equation:

$$100 : load = \tfrac{1}{4} : 10$$
$$load \times \tfrac{1}{4} = 10 \times 100$$
$$load = \frac{10 \times 100}{\tfrac{1}{4}} = 4000 \text{ lbs.}$$

The Wedge

This is virtually a pair of inclined planes in contact along their bases, or back to back. The wedge is generally driven by blows of a hammer or sledge instead of being pushed, as in the case of the other powers, although the wedge is sometimes moved by constant pressure. If the weight rests on a horizontal plane and a wedge is forced under it, the weight will be lifted a height equal to the thickness of the butt end of the wedge when the wedge has penetrated its length, as in Fig. 6-11.

Rule. As the applied force is to the load, so is the thickness of the wedge to its length. That is:

Applied force : load = thickness : length of wedge

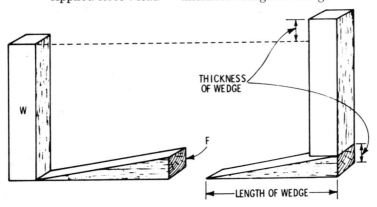

Fig. 6-11.   The application of the wedge in raising a heavy load.

Example—What force is necessary to apply to a wedge 20 in. long and 4 in. thick to raise a load of 2000 lbs.?

Substituting in the equation:

$$\text{Applied force} : 2000 = 4 : 20$$
$$20 : 4 = 2000 : \text{applied force}$$
$$\text{applied force} \times 20 = 4 \times 2000$$
$$\text{applied force} = \frac{4 \times 2000}{20} = 400 \text{ lbs.}$$

Expansion and Contraction

Practically all substances expand with an increase in temperature and contract with a decrease in temperature. The expansion of solid bodies in a longitudinal direction is known as *linear expansion*; the expansion in volume is called the *volumetric expansion*. A noticeable exception to the general law for expansion is the behavior of water. With a decrease in temperature, water will contract until it reaches its minimum volume, at a temperature of 39.1°F. This is the point of maximum density. With a continued decrease in temperature, the water will expand until it freezes and becomes ice, as shown in Fig. 6-12. Were it not for this fact, plumbers would be out of the job of repairing frozen pipes.

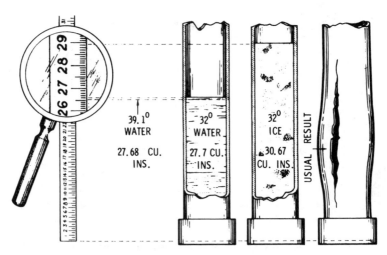

Fig. 6-12. The expansion of water at various temperatures.

Table 6-2. Linear Expansion of Common Metals (32°–212°F)

Metal	Linear expansion per unit length per degree F	Metal	Linear expansion per unit length per degree F
Aluminum	.00001234	Iron, wrought	.00000648
Antimony	.00000627	Lead	.00001571
Bismuth	.00000975	Nickel	.00000695
Brass	.00000957	Steel	.00000636
Bronze	.00000986	Tin	.00001163
Copper	.00000887	Zinc, cast ⎫	.00001407
Gold	.00000786	Zinc, rolled ⎭	
Iron, cast	.00000556		

Volumetric expansion = 3 × linear expansion.

The following example will illustrate the use of Table 6-2.

Example—How much longer is a 36-in. rod of aluminum when heated from 97° to 200°F?

$$\text{Increase in temperature is } 200 - 97 = 103°$$

$$\text{Coefficient of expansion for aluminum}$$
$$\text{from Table 6-2} = .00001234$$

$$\text{Increase in length of rod} =$$
$$36 \times .00001234 \times 103 = .0456 \text{ in.}$$

Melting Point of Solids

The temperatures at which a solid substance changes into a liquid is called the melting point. When a solid begins to melt, the temperature remains constant until the whole mass of the solid has changed into a liquid. The heat supplied during the period is used to change the substance from the solid to the liquid state and is called the *latent heat of fusion.*

For instance, to melt a pound of ice at 32°F requires 143.57 Btu, or 144 Btu for ordinary calculations. The temperature at which melting takes place varies for different substances, as shown in Table 6-3.

Impure metals usually have a lower melting point than pure metals. Low melting points may be obtained by combining several metals to form alloys. Often an alloy will melt at a much lower temperature than would be expected, considering the melting points

Table 6-3. Melting Points of Commercial Metals

Metal	Degrees F
Aluminum	1,200
Antimony	1,150
Bismuth	500
Brass	1,700 -1,850
Copper	1,940
Cadmium	610
Iron, cast	2,300
Iron, wrought	2,900
Lead	620
Mercury	−38
Steel	2,500
Tin	446
Zinc, cast	785

of the metals of which it is composed. Those of the lowest melting point contain bismuth, lead, tin, and cadmium.

By varying the percentages of each metal, melting points ranging from 149° to 324°F are obtained; these are only about one-fourth the melting point of the constituent metals. Alloys having such low fusing points are known as *low-fusing alloys*. These are considered further in Working with Copper Tubing, Chapter 2, in this volume, and Lead Work, Chapter 10, Volume I.

Gravity

By definition, gravity is *the force that attracts bodies, at or near the surface of the earth, toward the center of the earth*. This force varies at different points on the earth's surface. It is strongest at sea level, decreasing below sea level in the same ratio that its distance from the center of the earth decreases. Above the surface, the attraction decreases in ratio as the square of the distance from the center of the earth increases. Thus, a body weighs less on top of a high mountain than at sea level.

Falling Bodies

Under the influence of gravity *alone* all bodies fall to the earth with the same acceleration of velocity. Galileo proved this by dropping

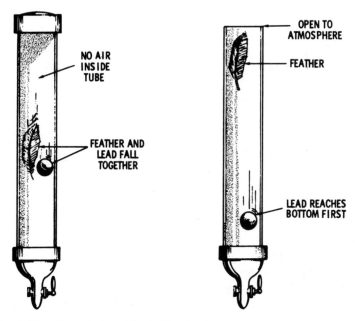

Fig. 6-13.   Experiment with falling bodies.

balls of different sizes at the same instant from the top of the leaning tower of Pisa. The spectators saw the balls start together and heard them strike the ground together. Of course, anybody knows that if, for instance, a feather and a piece of lead were released at the same time from an elevated point, the lead would reach the ground first. It is not the difference in weight that retards the feather but the effect of the air on the less dense object. In a vacuum, all bodies fall with the same acceleration of velocity, as has been proved by the experiment illustrated in Fig. 6-13.

Center of Gravity

Briefly, the center of gravity of a body is *that point of the body about which all its parts are balanced; or when that point is supported, the whole body will remain at rest though acted upon by gravity.* The center of gravity may be found by calculation and, in some cases, more conveniently by experiments, as in Fig. 6-14.

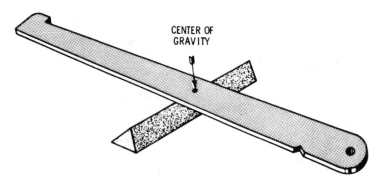

Fig. 6-14.   Method of finding the center of gravity.

Momentum

In popular language, momentum may be defined as *the power of overcoming resistance as possessed by a body by virtue of its motion*; that which makes a moving body hard to stop. Numerically, it is equal to the product of the mass of the body multiplied by its velocity. It is numerically equivalent to the number of pounds of force that will stop a moving body in 1 second, or the number of pounds of force which, acting during 1 second, will give it the given velocity.

Friction

By definition, friction is *that force which acts between two bodies at their surface of contact so as to resist their sliding on each other*; it is the resistance to motion when one body is moved upon another. Were it not for friction, many things would be impossible in mechanics. For instance, power could not be transmitted by belts, and automobiles could not be driven through clutches.

Because of friction, bearings must be lubricated, long pipe lines must be oversize to prevent undue loss of pressure, etc. The object of lubricating bearings is to form a film of oil so that the revolving part does not touch the bearing but revolves on the oil; the friction of solids on fluids is much less than that between solids. Ordinary bearings absorb from 3 percent to 5 percent of the applied power; roller bearings, 2 percent; ball bearings, 1 percent; spur gears with cast teeth, including bearings, 7 percent; spur gears with cut teeth,

4 percent; bevel gears with cast teeth, including bearings, 8 percent; bevel gears with cut teeth, including bearings, 5 percent; belting, 2 percent to 4 percent; roller chains, 3 percent to 5 percent.

Hydraulics

The term *hydraulics* is commonly, though ill-advisedly, defined as *the science that treats of liquids, especially water, in motion.* Properly speaking, there are two general divisions of the subject:

1. Hydrostatics.
2. Hydrodynamics.

Hydrostatics refers to liquids *at rest*, and hydrodynamics to liquids *in motion.* The outline given relates to water.

Water

Those who have had experience in the design or operation of pumps have found that water is an unyielding substance when confined in pipes and pump passages. This necessitates very substantial construction to withstand the pressure and periodic shocks or water hammer. Water at its maximum density (39.1°F) will expand as heat is added, and it will also expand slightly as the temperature falls from this point.

For ordinary calculations, the weight of 1 cu. ft. of water is taken at 62.4 lbs., which is correct when its temperature is 53°F. At 62°F the weight is 62.355 lbs. The weight of a U.S. gallon of water, or 231 cu. in., is roughly $8\frac{1}{3}$ lbs.

Head and Pressure

These are the two primary considerations in hydraulics. The word *head* signifies *the difference in level of water between two points,* and is usually expressed in feet. Two kinds of head are:

1. Static.
2. Dynamic.

The *static head* is the height from a given point of a column or body of water at rest, considered as causing or measuring pressure.

The *dynamic head* is an equivalent or virtual head of water in motion which represents the resulting pressure due to the height of the water from a given point, and the resistance to flow due to friction. Thus, when water is made to flow through pipes or nozzles, there is a loss of head. In ordinary calculations, it is common practice to estimate that every foot of head is equal to $\frac{1}{2}$ lb. of pressure per sq. in., as this allows for ordinary friction in pipes.

The following distinctions with reference to head should be carefully noted.

Total static head = static lift + static head.
Total dynamic head = dynamic lift + dynamic head.

Lift

When the barometer reads 30 in. at sea level, the pressure of the atmosphere at that elevation is 14.74 lbs. per sq. in. This pressure will maintain or balance a column of water 34.042 ft. high when the column is completely exhausted of air, and the water is at a temperature of 62°F. The pressure of the atmosphere then *lifts* the water to such a height as will establish equilibrium between the weight of the water and the pressure of the air. Similarly, in pump operation, the receding piston or plunger establishes the vacuum and the pressure of the atmosphere lifts the water from the level of the supply to the level of the pump. Accordingly, lift as related to pump operation may be defined as *the height in feet from the surface of the intake supply to the pump.*

Strictly speaking, lift is the height to which the water is elevated by atmospheric pressure, which in some pumps may be measured by the elevation of the inlet valve and in others by the elevation of the piston. The practical limit of lift is 20 to 25 ft. Long inlet lines, many inlet elbows, and high temperature of the water require shorter lifts. The lift must be reduced as the temperature of the water is increased, because the boiling point of water corresponds to the pressure.

Theoretically, a perfect pump will draw water from a height of 34 ft. when the barometer reads 30 in., but since a perfect vacuum cannot be obtained due to valve leakage, air entrained in the water, and the vapor of the water itself, the actual height is generally less than 30 ft., and for warm or hot water, considerably less. When the

Table 6-4. Theoretical Lift for Various Temperatures

Temp. F	Absolute pressure of vapor lbs. per sq. in.	Vacuum in inches of mercury	Lift in feet	Temp. F	Absolute pressure of vapor lbs. per sq. in.	Vacuum in inches of mercury	Lift in feet
102.1	1	27.88	31.6	182.9	8	13.63	15.4
126.3	2	25.85	29.3	188.3	9	11.6	13.0
141.6	3	23.83	27	193.2	10	9.56	10.8
153.1	4	21.78	24.7	197.8	11	7.52	8.5
162.3	5	19.74	22.3	202	12	5.49	6.2
170.1	6	17.70	20	205.9	13	3.45	3.9
176.9	7	15.67	17.7	209.6	14	1.41	1.6

water is warm the height to which it can be lifted decreases due to the increased pressure of the vapor. For example, a boiler feed pump taking water at 153°F could not produce a vacuum greater than 20.78 in. because, at that point, the water would begin to boil and fill the pump chamber with steam. Accordingly, the corresponding theoretical lift would be:

$$34 \times \frac{21.78}{30} = 24.68 \text{ ft., approximately}$$

The result is approximate because no correction has been made for the 34 which represents a 34-ft. column of water at 62°; of course, at 153° the length of such a column would be slightly increased. It should be noted that the figure 24.68 ft. is *approximate* theoretical lift for water at 153°; the *practical* lift would be considerably less. Table 6-4 shows the theoretical maximum lift for different temperatures, leakage not considered.

Flow of Water in Pipes

The quantity of water discharged through a pipe depends on:

1. The *head*, which is the vertical distance between the level surface of still water in the chamber at the entrance end of the pipe and the level of the center of the discharge end of the pipe.
2. The length of the pipe.

3. The smoothness of the interior surface of the pipe.
4. The number and sharpness of the bends. But the head is independent of the position of the pipe, as horizontal or inclined upward or downward. The head, instead of being an actual distance between levels, may be caused by pressure, as a pump, in which case the head is calculated as a vertical distance corresponding to the pressure. 1 lb. per sq. in. = 2.309 ft. head, or 1 ft. head = .433 lb. per sq. in.

Elementary Pumps

There are three elements necessary for the operation of a pump:

1. Inlet or suction valve.
2. Piston or plunger.
3. Discharge valve.

Simple pumps may be divided into two classes:

1. Lift pumps.
2. Force pumps.

A lift pump is one which does not elevate the water higher than the lift; a force pump operates against both lift and head.

Lift Pumps

Fig. 6-15 shows the essentials and working principle of a simple lift pump. In this type of pump there are two valves which are known as the *foot valve* and the *bucket valve*. During the upstroke, the bucket valve is closed and the foot valve is open, allowing the atmosphere to force the water into the cylinder. When the piston begins to descend, the foot valve closes and the bucket valve opens, which transfers the water in the cylinder from the lower side of the piston to the upper side.

During the next upstroke, the water already transferred to the upper side of the piston is discharged through the outlet. It will be noted that as the piston begins the upstroke of discharge it is subject to a small maximum head, and at the end of the upstroke to a minimum head. This variable head is so small in comparison to the

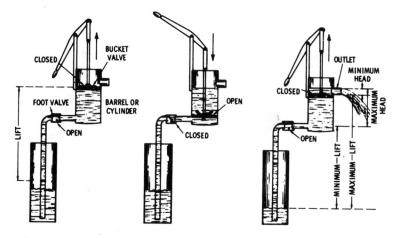

Fig. 6-15. Elementary single-acting lift pump showing the essential features and cycle of operation.

head against which a force pump works that it is not ordinarily considered.

Force Pumps

The essential feature of a force pump, which distinguishes it from a lift pump, is that *the cylinder is always closed*; in a lift pump it is *alternately closed and open* when the piston is respectively at the upper and lower ends of its stroke. In addition to the foot and bucket valves of the lift pump, a head valve is provided.

In operation, during the upstroke, atmospheric pressure forces water into the cylinder, and during the downstroke this water is transferred from the lower to the upper side of the piston. During the next upstroke the piston forces the water out of the cylinder through the head valve, which closes when the piston reaches the end of the stroke and the cycle is repeated.

A simple form of force pump is one known as a single-acting plunger pump. In operation, during the upstroke, water fills the cylinder, the inlet valve opens, and the outlet valve closes. During the downstroke, the plunger displaces the water in the barrel, forcing it through the discharge valve against the pressure head.

A piston is *shorter* than the stroke, whereas a plunger is *longer*

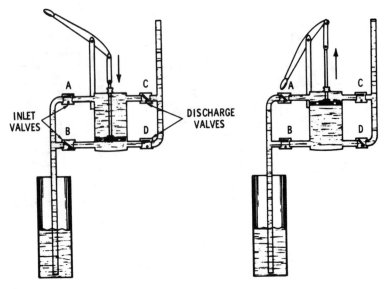

Fig. 6-16.   Elementary double-acting force pump.

than the stroke. The word plunger is very frequently used erroneously for a piston.

Double-Acting Force Pump

By fitting a set of inlet and outlet valves at each end of a pump
cylinder, it is made double-acting; that is, a cylinder full of water
is pumped each stroke instead of every other stroke. With this
arrangement, the piston can have approximately half the area of the
single-acting piston for equal displacement. Accordingly, the maximum stresses brought on the reciprocating parts are reduced approximately one-half, thus permitting lighter and more compact
construction. In the double-acting pump there are no bucket valves,
a solid piston being used. The essential features and operation are
plainly shown in Fig. 6-16. Three are two inlet valves, A and B,
and two discharge valves, C and D, the cylinder being closed and
provided with a piston. In operation, during the downstroke, water
follows the upper face of the piston through valve A. At the same
time, the previous charge is forced out of the cylinder through valve

D by the lower face of the piston. During these simultaneous operations, valves *A* and *D* remain open, and *B* and *C* are closed.

During the upstroke, water follows the lower face of the piston through valve *B*. At the same time, the previous charge is forced out of the cylinder through valve *C* by the upper face of the piston. During these simultaneous operations, valves *B* and *C* remain open, and *A* and *D* are closed.

CHAPTER 7

Grooved Piping Systems

Grooved piping systems offer an alternative to threaded, welded, and flanged piping systems. When grooved systems are installed, each joint becomes a union, permitting easy access to any part of a system for cleaning or servicing. Flexible couplings can be used to provide for expansion, contraction, or deflection when needed. Flexible couplings may eliminate need for expansion joints or loops and provide a virtually stress-free piping system. Rigid couplings may be used to provide positive clamping for flexural and torsional loads, e.g., valve connections, machinery rooms, fire mains, and long, straight runs of piping. When grooved-joint piping is used, the slight gap between pipe ends isolates noise and vibration, as shown in Fig. 7-1. A roll grooving machine for pipe shop use is shown in Fig. 7-2. The machine in Fig. 7-3 is designed for on-the-job use.

Roll grooving is applicable for steel, stainless, aluminum, and similar metallic pipe, as well as up to Schedule 80 PVC. A complete range of pipe fittings is available for use with grooved piping systems. These include but are not limited to:

Elbows: short pattern, long pattern, reducing, 90°, 45°, $22\frac{1}{2}°$, and $11\frac{1}{4}°$.

Tees: straight, reducing, crosses.

Couplings: straight, reducing, quick-disconnect, threaded or grooved outlet.

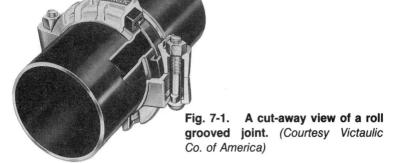

Fig. 7-1.   A cut-away view of a roll grooved joint. *(Courtesy Victaulic Co. of America)*

Fig. 7-2.   This grooving machine is designed for shop use. *(Courtesy Victaulic Co. of America)*

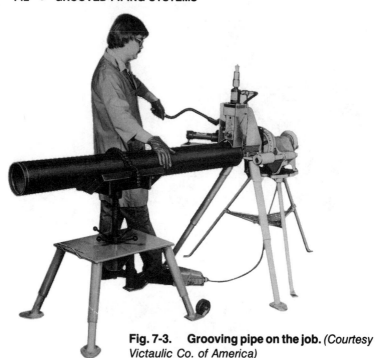

Fig. 7-3.   Grooving pipe on the job. *(Courtesy Victaulic Co. of America)*

Wyes: straight, reducing, long turn tee-wye.
Flanges: flange adaptors.
Caps.
Reducers: grooved, threaded, concentric, eccentric, swaged.
Sprinklers: head fittings.
Strainers.

Elbows and tees, Fig. 7-4, are among the many fittings used with grooved piping systems. A grooved flange is shown in Fig. 7-5.

Valves

Many types of valves are available for use with grooved pipe systems. These include but are not limited to:

Butterfly valves: lever, geared, actuated, two-way, three-way.
Check valves: disc, dual disc, swing.
Ball valves.

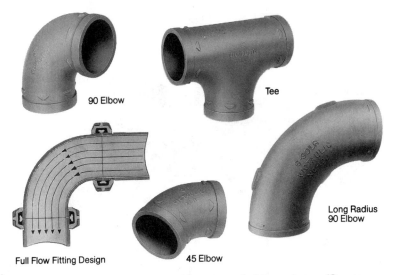

Fig. 7-4. Elbows and tees used in grooved pipe systems. *(Courtesy Victaulic Co. of America)*

Fig. 7-5. Installing a flange on a grooved joint. *(Courtesy Victaulic Co. of America)*

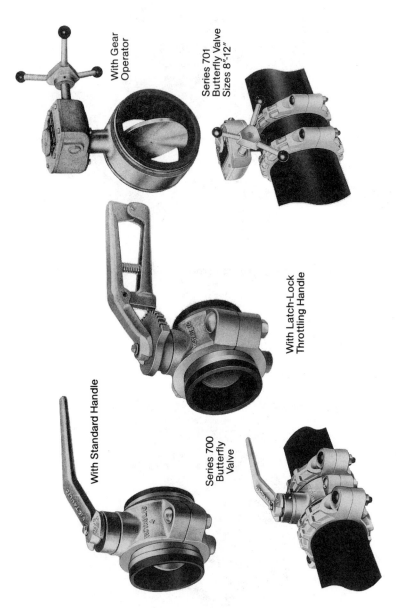

With Gear
Operator

Series 701
Butterfly Valve
Sizes 8"-12"

With Latch-Lock
Throttling Handle

With Standard Handle

Series 700
Butterfly
Valve

Fig. 7-6. Various types of butterfly valves. *(Courtesy Victaulic Co. of America)*

Fig. 7-7. **A dual-disc check valve.** *(Courtesy Victaulic Co. of America)*

Various types of butterfly valves are shown in Fig. 7-6. A dual-disc check valve is shown in Fig. 7-7.

Applications of Grooved Piping

General Construction

Grooved piping systems are used in buildings for:

> Domestic hot and cold water.
> Storm and roof drains.
> Hydronic heating systems.
> Underground services: water mains, fire mains.
> Chilled water.
> Air-conditioning systems.
> Cooling tower piping.
> Condenser water.

Oil Field Piping

Grooved piping systems are routinely used in oil field services for:

Well service lines.
Pipeline headers.
Flow lines.
Mud lines.
Injection lines.
Salt water disposal.
Water flood plants.
Production headers.
Wellhead hookup.
Gathering lines.

Military Services

Rapid deployment piping.
Mobile equipment piping.
Disaster control piping.

Plus other services where speed of installation is very important.

Fire Protection Systems

Wet and dry standpipes.
Automatic sprinklers.
Feed mains.
Cross mains.
Branch lines.
Underground services.

Grooved piping systems are particularly adapted to fire protection systems. Grooved pipe can be delivered to the job site pre-cut to length, and if job conditions force slight on-the-job alterations, on-the-site grooving and the availability of needed fittings make changes easy.

Two types of grooving are used in grooved piping systems:

Roll Grooving and Cut Grooving

Roll Grooving

The concept of cold forming (rolling) a groove into pipe was first used to groove pipe with insufficient wall thickness (light wall) to permit cut grooving. Further development of the roll grooving process led to successful roll grooving of standard weight steel pipe. Victaulic roll grooving tools, Figs. 7-2 and 7-3, are designed to rotate the pipe as an upper roll (the male die) is impressed into the pipe. The lower roll, in addition to driving the pipe, is the female die inside the pipe. The groove depth is controlled by an adjustable stop.

The roll grooving process removes no metal, displacing metal to form the groove. Roll grooving is applicable for steel, stainless steel, aluminum, and similar pipe and can be used on up to Schedule 80 PVC pipe. The machines shown in Figs. 7-2 and 7-3 can be used to roll groove pipe up to 16-inch.

Cut Grooving

Cut grooving differs from roll grooving in that a groove is cut into the pipe. Cut grooving is basically intended for standard weight or heavier pipe. The cut removes less than one-half the pipe wall, which is less depth than thread cuts. Cut grooving machines are designed to be driven around a stationary pipe. This assures a groove which is concentric with the pipe O.D. (outside diameter) and is of uniform depth. The cut grooving tool, Fig. 7-8, is driven by an external power source. The tool shown in Fig. 7-9 is designed for either manual or power cut grooving.

Regardless of whether the roll grooving or the cut grooving method is used, a variety of couplings and gaskets are available to suit the intended service. Flexible couplings, Fig. 7-10A, provide allowances for controlled pipe movement, expansion, contraction, deflection, absorption of movement from thermal changes, settling or seismic action, as well as dampening noise and vibration. Rigid couplings, Fig. 7-10B, create a rigid joint, useful for risers, mechanical rooms, and other areas where rigid piping is desirable.

Fig. 7-8. **An external-powered cut groover.** *(Courtesy Victaulic Co. of America)*

Fig. 7-9. **This cut groover is designed for either external power or manual operation.** *(Courtesy Victaulic Co. of America)*

FLEXIBLE

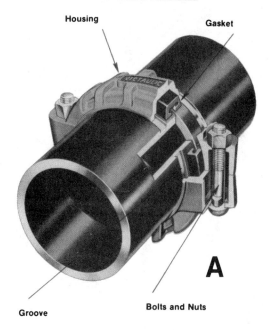

Housing

Gasket

A

Groove

Bolts and Nuts

RIGID

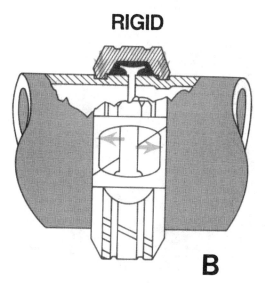

B

Fig. 7-10. **Flexible and rigid couplings.** *(Courtesy Victaulic Co. of America)*

Fig. 7-11.   Grooved piping installed in oil fields. *(Courtesy Victaulic Co. of America)*

Fig. 7-12.   Grooved piping in a waste-water treatment plant. *(Courtesy Victaulic Co. of America)*

Grooved pipe installations are used extensively in oil fields and oil tank storage areas as shown in Fig. 7-11. The ease of accessibility to any part of a piping system is shown in Fig. 7-12, part of the sludge treatment piping in a waste-water treatment plant.

CHAPTER 8

General Plumbing Information

The ability to perform basic calculations is a prime necessity, not only to get a plumber's license, but for the master plumber as well. With this in mind, a number of practical examples, or problems bearing on the subject matter, have been calculated. Although requirements for a plumber's license differ in various localities, the problems given are of a type usually found in license examinations for master plumbers. In this connection, it should be impressed upon all candidates for licenses the great necessity of study to master the fundamental principles underlying each example given, in order to increase abilities and be able to solve any new or similar problem at the written examination.

Example—A water tank is 65 in. long and has a trapezoidal cross section. The two parallel sides are 32 in. and 44 in., and the distance between them is 40 in. Determine the capacity of the tank in gallons.

Solution—The formula for calculating the area of a trapezoid is:

$$A = \frac{1}{2} H (a + b)$$

where

H is the distance between the two parallel sides

a and b are the lengths of the parallel sides

A is the area

A substitution of numerical values gives:

$$A = \frac{1}{2} \times 40 \ (32 + 44) = 1520 \text{ sq. in.}$$

The volume in cubic inches = 1520 × 65 = 98,800

Since there are 231 cu. in. in 1 gallon, the gallon contents of the tank =

$$\frac{98,800}{231} = 427.7 \text{ gals.}$$

Example—What is the radius of a circle the area of which is equal to that of a rectangle whose sides are 26.9 and 12.5 in. respectively?

Solution—Area of rectangle = 26.9 × 12.5 = 336.25 sq. in.

Area of circle = πR^2 = 336.25 sq. in.

From which R = $\sqrt{\dfrac{336.25}{\pi}}$ = 10.34 in.

Example—What is the weight of a solid ball of brass 6 in. in diameter? Assume specific gravity = 8.4.

Solution—The cubic contents of the ball is obtained by the use of the following formula:

$$V = \frac{4\pi R^2}{3} = \frac{4\pi \times 3^3}{3} = 113.1 \text{ cu. in.}$$

Weight of the ball equals the weight of 1 cu. in. of water × specific gravity of brass × circumference of the ball. Therefore:

Weight of ball = 0.0361 × 8.4 × 113.1 = 34.3 lbs.

Example—Find the height of a cast-iron cone whose weight is 533.4 kilograms, and whose diameter at the base is 5 decimeters. (Specific gravity of cast iron = 7.22.)

Solution—The cubic content of the cone is:

$$\frac{533.4}{7.22} = 73.9 \text{ cubic decimeters}$$

$$\text{Volume of cone} = \frac{\text{base area} \times H}{3}$$

It follows that

$$73.9 = \frac{5 \times 5 \times 0.7854 \times H}{3}$$

and

$$H = \frac{3 \times 73.9}{5 \times 5 \times 0.7854}$$
$$= 11.3 \text{ decimeters, or } 11.3 \times 3.937$$
$$= 44.5 \text{ in. (approx.)}$$

Example—In a certain plumbing installation, three pipes have an internal diameter of 2, $2\frac{1}{2}$, and 3 in., respectively. What is the diameter of a pipe having an area equal to the three pipes?

Solution—The areas of the three pipes are as follows:

$$A_1 + A_2 + A_3 \times \frac{\pi \times 2^2}{4} + \frac{\pi \times 2.5^2}{4} + \frac{\pi \times 3^2}{4}$$

$$= \frac{\pi}{4} (2^2 + 2.5^2 + 3^2)$$

$$= \frac{\pi}{4} (4 + 6.25 + 9)$$

$$= \frac{\pi}{4} \times 19.25 \text{ sq. in.}$$

Since the formula for a circular area is $\frac{\pi \times D^2}{4}$ we obtain,

$$\frac{\pi}{4} \times 19.25 = \frac{\pi}{4} \times D^2, \text{ or } D^2 = 19.25 \text{ sq. in.}$$

Therefore,

$$D = \sqrt{19.25}, \text{ or } 4.39 \text{ in.}$$

Remember that the area of any circular pipe is directly proportional to the square of its diameter in inches, so our calculation will be somewhat simplified. We have:

$$2^2 + 2.5^2 + 3^2 = D^2$$

or,

$$D = \sqrt{19.25} = 4.39 \text{ in.}$$

It follows from the foregoing that the area of a pipe having the same capacity as a 2-, $2\frac{1}{2}$-, and 3-in.-diameter pipe together must be 4.39 in. in diameter, or $4\frac{25}{64}$ in.

Example—If a 1-in. pipe delivers 10 gal. of water per minute, what size of pipe will be required to deliver 20 gal. per minute?

Solution—In this problem, it is necessary to find the diameter of a piece of pipe whose area is twice as large as one that is 1 inch in diameter.

Since the area for any circular section $= \dfrac{\pi D^2}{4}$, it follows that the area for the 1-in. pipe is $\dfrac{\pi \times 1^2}{4} = 0.7854$ sq. in. A pipe having twice this area consequently has a diameter of $\dfrac{\pi}{2}$. Thus, $\dfrac{\pi}{2} = \dfrac{\pi D^2}{4}$, which after rearrangement of terms gives $D^2 = 2$, or $D = \sqrt{2} = 1.4142$ in. That is, the required diameter of a pipe to deliver 20 gpm is $1\frac{13}{32}$ in. (approximately).

Example—If the total fall of a house sewer is 24 in. per 120 ft., what is the slope per foot of this sewer?

Solution—Since the sewer has a total length of 120 ft., the slope per foot is $\frac{24}{120}$, or 0.2 in.

Example—What is the weight of 1 cu. in. of water if it is assumed that 1 cu. ft. weighs 62.5 lbs.?

Solution—1 cu. ft. contains 12 × 12 × 12, or 1728 cu. in. Therefore, 1 cu. in. weighs $\frac{62.5}{1728}$, or 0.0361 lb. This figure is given in most handbooks and is frequently used to calculate water pressure in tanks, pipes, etc.

Example—What is the weight of a column of water 12 in. high and 1 in. in diameter?

Solution—Since 1 cu. in. of water weighs 0.0361 lb., a column of water 12 in. high weighs 12 × 0.0361, or 0.4332 lb.

Example—What is the pressure in pounds per square inch (psi) at the base of a 10-ft. water cylinder?

Solution—The pressure may be found by remembering that the 10-ft. water column weighs 10 × 12 × 0.0361, or 4.332 psi.

Example—Calculate the minimum water pressure at the city water main, in pounds per square inch (psi), necessary to fill a house tank located on top of a 6-story building when the inlet to the water tank is located at an elevation of 110 ft. above the city water main.

Solution—Since 1 cu. in. of water weighs 0.0361 lb., the minimum pressure required is 12 × 110 × 0.0361, or 47.7 psi.

Example—A water tank 10 ft. high and 15 ft. across is to be constructed of wood. When filled to the bottom of the overflow pipe, its capacity is 11,000 gals. How high above the inside bottom of the tank should the bottom of the overflow pipe be in order to have the required water capacity in the tank? (One cubic foot equals 7.48 gals.).

Solution—The area of the tank in square feet multiplied by the assumed height of the water in the tank in feet equals the cubic content of the water in cubic feet. Since 1 cu. ft. equals 7.48 gals., the cubic content occupied by the water is $\frac{11000}{7.48}$, or 1470.6 cu. ft.

The equation required for our calculation will therefore be as follows:

$$7.5^2 \times \pi \times H = 1470.6$$

$$H = \frac{1470.6}{56.25\pi} = 8.32 \text{ ft.}$$

Thus, the bottom of the overflow pipe should be located 8.32 ft. or 8 ft. $3\frac{27}{32}$ in. above the bottom of the tank.

Example—What is the number of horsepower required to raise 40,000 lbs. 200 ft. in 5 minutes? Disregard losses.

Solution—By definition, 1 horsepower is equivalent to doing work at a rate of 33,000 lbs. per minute. Thus, in the present problem

$$HP = \frac{\text{foot-pounds}}{33,000 \times t}$$

where

HP is the horsepower required
t is the time in minutes

Substituting values, we obtain

$$HP = \frac{\text{foot-pounds}}{33,000 \times t} = \frac{40,000 \times 200}{33,000 \times 5} = 48.5 \text{ HP}$$

Example—What is the number of horsepower required to lift 10,000 gals. of water per hour to a height of 90 ft.? Disregard losses. (Assume weight of water to be $8\frac{1}{3}$ lbs. per gallon.)

Solution:

$$HP = \frac{\text{foot-pounds}}{33,000 \times t} = \frac{10,000 \times 8\frac{1}{3} \times 90}{33,000 \times 60} = 3.79 \text{ HP}$$

Example—A city of 25,000 uses 15 gals. of water per day per capita. If it is required to raise this water 150 ft., what is the number of horsepower required? Disregard losses.

Solution:

$$HP = \frac{\text{foot-pounds}}{33,000 \times t}$$

$$= \frac{15 \times 8\frac{1}{3} \times 25,000 \times 150}{33,000 \times 60 \times 24} = HP \text{ (approx.)}$$

Example—How many gallons of water can a 75-HP engine raise 150 ft. high in 5 hours? One gallon of water weighs $8\frac{1}{3}$ lbs. Disregard losses.

Solution—If the given data is substituted in our formula for horsepower, we obtain:

$$75 = \frac{150 \times 8\frac{1}{3} \times G}{33,000 \times 5 \times 60}$$

$$G = \frac{75 \times 33,000 \times 5 \times 60}{150 \times 8\frac{1}{3}} = 594,000 \text{ gals.}$$

Example—A circular tank 20 ft. deep and 20 ft. in diameter is filled with water. If the average height to which the water is to be lifted is 50 ft., what must be the horsepower of an engine capable of pumping the water out in 2 hours? Disregard losses.

Solution—In this example, it is first necessary to calculate the cubic content of the tank; that is, the cross-sectional area multiplied by its height. Thus,

volume in cu. ft. $= \pi R^2 H = \pi \times 10^2 \times 20 + 6283$ cu. ft.

Since 1 cu. ft. of water weighs 62.5 lbs., the total weight of the tank's contents is:

$$62.5 \times 6283 = 392,700 \text{ lbs. (approx.)}$$

Again, using our formula, we have:

$$HP = \frac{\text{foot-pounds}}{33,000 \times t} = \frac{392,700 \times 5}{33,000 \times 2 \times 60} = 5 \text{ HP (approx.)}$$

Example—The suction lift on a pump is 10 ft. and the head pumped against is 100 ft. If the loss due to friction in the pipe line is assumed as 9 ft., and the pump delivers 100 gals. per minute, what is the horsepower delivered by the pump?

Solution—When the water delivered is expressed in gallons per minute (usually written gpm) the formula for horsepower is:

$$HP = \frac{gpm \times \text{head in feet} \times 8.33}{33,000}$$

A substitution of values gives

$$HP = \frac{100 \times 119 \times 8.33}{33,000} = 3\,HP$$

Example—A tank having a capacity of 10,000 gals. must be emptied in 2 hours. What capacity pump is required?

Solution—The capacity of the pump in gallons per minute (gpm) is arrived at by dividing the total gallonage of the tank by the time in minutes. Thus,

$$gpm = \frac{10,000}{2 \times 60} = 83.3\,gpm$$

Example—What is the net capacity of a double-acting pump having a piston diameter of 3 in. and a stroke of 5 in. when it makes 75 strokes per minute? Assume slip of pump = 5 percent.

Solution—The rule for obtaining pump capacity is as follows: *Multiply the area of the piston in square inches by the length of the stroke in inches, and by the number of delivery strokes per minute; divide the product by 231 to obtain the theoretical capacity in U.S. gallons.*

This rule is most commonly stated in a formula as follows:

$$gpm = \frac{D^2 \times 0.7854 \times L \times N}{231}$$

where,

      gpm = number of gallons pumped per minute.
        D = diameter of plunger or piston in inches.
        L = length of stroke in inches.
        N = number of delivery strokes per minute.

A substitution of values in the above formula gives:

$$\text{gpm} = \frac{3^2 \times 0.7854 \times 5 \times 75}{231} = 11.5$$

With a slip of 5%, the total net capacity of the pump is finally $11.5 \times 0.95 = 10.9$ gpm.

Example—What is the hourly net capacity of a 2×8 double-acting power pump running at 150 rpm and having a slip of 10 percent?

Solution—The formula for pump capacity is:

$$\text{gpm} = \frac{D^2 \times 0.7854 \times L \times N}{231}$$
$$= \frac{2^2 \times 0.7854 \times 8 \times 300}{231} = 32.7 \text{ gpm}$$

The hourly net capacity $= 32.7 \times 60 \times 0.9 = 1765.8$ gals.

It should be observed that if N is taken to represent the number of revolutions of a single double-acting pump, the result is to be multiplied by 2, and if N represents the number of revolutions of a duplex pump, which would be the same as two single pumps, the result must be multiplied by 4.

Example—What is the capacity of a double-acting pump in gallons per minute if the cylinder is 9 in. in diameter, and the stroke 10 in., when it makes 60 strokes per minute? (Disregard slippage.)

Solution—The capacity of the pump in gallons per minute (gpm) is:

$$\text{gpm} = \frac{D^2 \times 0.8754 \times L \times N}{231}$$
$$= \frac{9^2 \times 0.7854 \times 10 \times 60}{231}$$
$$= 165.2 \text{ gpm}$$

Example—A gas engine has a 4-in. piston and the effective pressure acting upon it is 50 lbs. per sq. in. What is the total load on the piston?

Solution—In this example, it is first necessary to determine the total net area of the piston. In multiplying this area by the pressure, we obtain the total load acting upon the piston. Thus,

$$A = \frac{\pi \times 4^2}{4} = 12.57 \text{ sq. in.}$$

$$\text{Total load} = 12.57 \times 50 = 628.5 \text{ lbs.}$$

Example—What is the indicated horsepower of a 4-cylinder 5 × 6 engine running at 500 rpm and 50 lbs. per sq. in. effective pressure?

Solution—The well-known formula for calculation of indicated horsepower is:

$$\text{IHP} = \frac{\text{PLAN}}{33,000} \times \text{K}$$

where coefficient (K) = 2 (four-cylinder engine). Substituting our values, we obtain:

$$\text{IHP} = \frac{50 \times \dfrac{6}{12} \times 0.7854 \times 5^2 \times 500}{33,000} \times 2 = 14.87$$

Example—The temperature of a furnace as registered by a pyrometer is 2750°F. What is the corresponding reading in Centigrade degrees?

Solution—The equation is:

$$C = \tfrac{5}{9}(F - 32)$$

$$C = \tfrac{5}{9}(2750 - 32) = 1510°$$

Example—A pail contains 58 lbs. of water having a temperature of 40°F. If heat is applied until the temperature of the water reaches 95°F, what is the amount of Btu supplied to the water?

Solution—The rise in temperature has been 95 − 40 = 55°F. Since one Btu is one pound raised one degree, it follows that to raise 58 pounds 55 degrees requires:

$$58 \times 55 = 3190 \text{ Btu}$$

Example—How many heat units (Btu) are required to raise one pound of water from 55° to 212° F? How many units of work does this represent?

Solution—The number of heat units required is 1 (212 − 55) = 157 Btu.

Since the mechanical equivalent of heat is 778, it is only necessary to multiply the number of heat units by this constant to obtain the equivalent number of work units. Thus,

$$157 \times 778 = 122,146 \text{ foot-pounds}$$

Example—In a certain pump installation, it was found that a 2-in. pipe, due to corrosion, had an effective diameter of only 1½ in. Calculate the loss in cross-sectional area due to corrosion.

Solution—It may easily be shown that the area of any circular pipe varies as the square of its diameter. The loss in cross-sectional area is therefore:

$$(2 \times 2) - (1.5 \times 1.5) = 4 - 2.25, \text{ or } 1.75$$

$$1.75 \times .7854 = 1.37 \text{ sq. in. loss}$$

(Note: .7854 = area for a 1-in. pipe.)

Example—It is required to calculate the pipe size for a shallow-well suction pump having a capacity of 300 gals. per hour (see Fig. 8-1). The horizontal distance between the pump and the well is 75 ft. and the vertical lift is 20 ft. plus the 5 ft. below water level, as shown in the illustration.

Solution—Bearing in mind that 22 ft. is considered the maximum practical suction lift for a shallow-well pump, our calculation will be as follows:

Total water lift =   20 ft.
Total pipe friction loss assuming 5-gpm
    flow through a 1-in. pipe.

Vertical part of pipe	= 25 ft.
Horizontal part of pipe	= 75 ft.
1-in. 90° elbow	= 6 ft.
Total footage	= 106 ft.

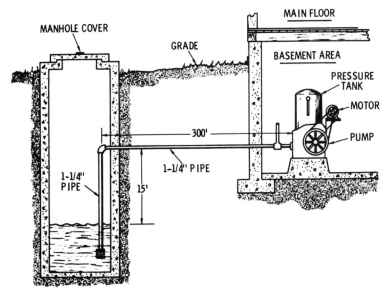

Fig. 8-1. Piping arrangement for a shallow-well suction pump.

As noted in Table 8-1, the friction loss in a 1-in. pipe at 5 gals. per minute flow equals 3.25 ft. per 100 ft. of pipe. Since there are 106 ft., the total friction loss = 106/100 × 3.25 = 1.06 × 3.25 = 3.4 ft. The total lift is 20 + 3.4, or 23.4 ft. As noted from our figures, a total lift of 23.4 ft. exceeds 22 ft. by a considerable margin, and

Table 8-1. Friction Loss in Pipe

Flow	Size of Pipe											
Gals.	½ inch		¾ inch		1 inch		1¼ inch		1½ inch		2 inch	
per Min.	Ft.	Lbs.	Ft.	Lbs.	Ft.	Lbs.	Ft.	Lbs.	Ft.	Lbs.	Ft.	Lbs.
2	7.4	3.2	1.9	.82								
3	15.8	6.85	4.1	1.78	1.26	.55						
4	27.0	11.7	7.0	3.04	2.14	.93	.57	.25	.26	.11		
5	41.0	17.8	10.5	4.56	3.25	1.41	.84	.36	.40	.17		
6			14.7	6.36	4.55	1.97	1.20	.52	.56	.24	.20	.086
8			25.0	10.8	7.8	3.38	2.03	.88	.95	.41	.33	.143
10			38.0	16.4	11.7	5.07	3.05	1.32	1.43	.62	.50	.216
12					16.4	7.10	4.3	1.86	2.01	.87	.70	.303
14					22.0	9.52	5.7	2.46	2.68	1.16	.94	.406
16					28.0	12.10	7.3	3.16	3.41	1.47	1.20	.520
18							9.1	3.94	4.24	1.83	1.49	.645

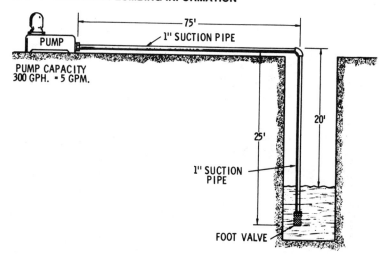

Fig. 8-2. Piping arrangement for a shallow-well basement pump.

although it is possible that an installation such as the foregoing will work, it should not be attempted as a practical solution.

If, on the other hand, a $1\frac{1}{4}$-in. pipe is selected, a similar reference to our friction-loss table indicates that the friction loss is reduced to 0.84 ft., making a total suction lift of 1.06 × 0.84 + 20 = 20.9 ft., which is within the practical suction limit.

Example—It is necessary to install a shallow-well basement pumping system (see Fig. 8-2). The horizontal distance between pump and well is 300 ft. and the vertical lift is 15 ft. Determine the pipe size and pressure-switch setting for a 5-gpm pump.

Solution—With reference to our friction loss (Table 8-1), it will be noted that, if a 1-in. pipe is selected, a 5-gpm flow will cause a friction loss of 3.25 ft. per 100 ft. Since the total pipe length equals 321 ft. (which figure includes the friction loss through the foot valve and elbow), we obtain a total friction loss of 3.25 × 3.21, or 10.43 ft. The total suction lift is therefore 15 + 10.43, or 25.43 ft. If, on the other hand, a $1\frac{1}{4}$-in. pipe is selected, a similar reference to the friction-loss table will give a total loss of 0.84 × 3.21, or 2.7 ft. The total suction lift using the larger pipe will be 15 + 2.7, or 17.7 ft., which is the size that should be used for this particular installation.

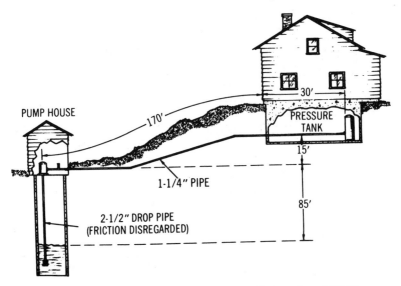

Fig. 8-3. Piping arrangement for a deep-well pumping system.

In order to calculate the necessary pressure to overcome the 15-ft. elevation and pipe friction loss, the total suction lift value of 17.7 must be multiplied by 62.5/144 or 0.434; that is, 17.7 × 0.434 equals 7.7 lbs. pressure. Pressure switches are usually set at 20 lbs. minimum and 40 lbs. maximum pressure. If 7.7 lbs. is added to the foregoing, we will arrive at a minimum switch setting of 27.7 and a maximum of 47.7 lbs.

Example—Water is to be pumped to a pressure tank (Fig. 8-3) in the basement of a home by an electric motor from a well in which the surface of the water is 85 ft. below the pump head. The tank is 15 ft. higher than the pump. Maximum pressure in the tank is 40 lbs. The distance from the well to the house is 170 ft., and 30 ft. of pipe is required in the house to reach the pressure tank. A $2\frac{1}{2}$-in. drop pipe is being used between the well and the pump, and a $1\frac{1}{4}$-in. pipe from pump to tank. If it is assumed that a 480-gal.-per-hour pump is used in the installation, what size motor will be required?

Solution—In order to establish the motor size, it will first be necessary to calculate the total head of the pump installation. From the foregoing data we obtain:

Head due to difference in elevation $= 2 \times 2.03 = 4.06$ ft.
Head due to pressure at tank $= 85 + 15 = 100.00$ ft.
Total pipe friction loss for the 200
ft. of $1\frac{1}{4}$-in. pipe, assuming
8 gpm flow (from table) $= 40 \times 2.31 = \underline{92.40 \text{ ft.}}$
Total head $= 196.46$ ft.
(200 ft. approx.)

The theoretical horsepower required may be determined by multiplying the gallons per minute by the total head in feet and dividing the product by 33,000. Since the pump efficiency is not known, we may assume an arbitrary value for our deep-well pump as 30 percent.

The actual horsepower, therefore $= \dfrac{8 \times 8.34 \times 200}{33,000 \times 0.3} = 1\frac{1}{3}$

Use the next larger standard-size electric motor, which is $1\frac{1}{2}$ horsepower.

Example—A double-acting single-piston pump has a $2\frac{1}{2}$-in.-diameter cylinder and a 3-in. stroke. What is the capacity per revolution?

Solution—In a problem of this type, it will first be necessary to calculate the piston area, which is $2.5^2 \times 0.7854$, or 4.909 sq. in. The pump capacity per stroke for one single-acting cylinder in gallons is obviously $4.91 \times 3/231$, or 0.064. Since there is one forward-and-back stroke per revolution in a double-acting pump, the pump capacity per revolution is 0.064×2, or 0.128 gal.

Effect of Pipe Friction Loss

The friction loss in piping is a very important factor, and must be taken into account when evaluating a water distribution system. The friction loss shown in Table 8-1 is based on a section of 15-year-old pipe. With reference to this table, it is easy to determine the friction loss through any one of the pipe sizes shown for any flow of water.

Thus, for example, a check in our friction-loss table indicates that discharge rate of 5 gals. per minute (gpm) through 100 ft. of 1-in. iron pipe results in a friction loss of 3.25 ft. The same gpm through 100 ft. of $\frac{3}{4}$-in. pipe will result in a friction loss of 10.5 ft. From this, it will be noted that pipe friction must be taken into consideration when pipe is selected for the suction line on a shallow-well pump, or for the discharge pipe from the pressure tank to the point of delivery.

Pump Characteristics

Automatic and semiautomatic pump installations for water supply purposes commonly employ three types of pumps. While each type has its individual characteristics, they all conform to the same general principles. There is the *reciprocating* or *plunger type*, the *rotary*, and the *centrifugal*; the *jet* or *ejector* pump has derived its name from the introduction of a jet system attached to the centrifugal type of pump. A brief tabulation of the characteristics of the various types of pumps is given in Table 8-2. For convenience, they are listed as to speed, suction lift, and practical pressure head.

In the selection of pumps, it cannot be too strongly emphasized that since each pump application will differ, not only in capacity requirement but also in the pressure against which the pump will have to operate, plus other factors, the pump manufacturer should be consulted as to the type of pump that will be best suited for a particular installation.

For example: If a $\frac{1}{2}$-in. hose with nozzle is to be used for sprinkling, water will be consumed at the rate of 200 gals. per hour. To permit use of water for other purposes at the same time, it is therefore essential to have a pump capacity in excess of 200 gals. per hour. Where $\frac{1}{2}$-in. hose with nozzle is to be used, we recommend the use of a pump having a capacity of at least 220 gals. per hour, which leaves available, for other uses, water at the rate of 20 gals. per hour when the hose is being used. In determining the desired pump capacity, even for ordinary requirements, it is advisable to select a size large enough so that the pump will not run more than a few hours per day at the most.

Table 8-2. Pump Characteristics

Type Pump	Speed	Practical Suction Lift	Pressure Head	Delivery Characteristics
Reciprocating: Shallow Well (low pressure (medium pressure)	Slow 250 to 550 strokes per min.	22 to 25 ft.	40 to 43 lbs. Up to 100 lbs.	Pulsating (air chamber evens pulsations) "
(high pressure) Deep Well	Slow 30 to 52 strokes per min.	Available for lifts up to 875 ft. Suction lift below cylinder 22 ft.	Up to 350 lbs. Normal 40 lbs.	" "
Rotary Pump: (shallow well)	400 to 1725 rpm	22 ft.	About 100 lbs.	Positive (slightly pulsating)
Ejector Pump: (shallow well and limited deep wells)	Used with centrifugal-turbine or shallow well reciprocating pump	Max. around 120 ft. Practical at lifts of 80 ft. or less	40 lbs. (normal) Available at up to 70 lbs. pressure head	Continuous nonpulsating, high capacity with low-pressure head
Centrifugal: Shallow Well (single stage)	High, 1750 and 3600 rpm	15 ft. maximum	40 lbs. (normal) 70 lbs. (maximum)	Continuous nonpulsating, high capacity with low-pressure head
Turbine Type: (single impeller)	High, 1750 rpm	28 ft. maximum at sea level	40 lbs. (normal) Available up to 100 lbs. pressure head	Continuous nonpulsating, high capacity with low-pressure head

Reciprocating pumps will deliver water in quantities proportional to the number of strokes and the length and size of the cylinder. They are adapted to a wide range of speeds, and to practically any depth of well. Since reciprocating pumps are positive in operation, they should be fitted with automatic relief valves to prevent rupture of pipes or other damage, should power be applied against abnormal pressure.

If it is not practical to set the pump directly over the well, as is necessary with deep-well plunger pumps, an ejector-type pump may be selected. The ejector pump is most efficient where the lift is between 25 and 65 ft., but will operate with lifts of up to 120 ft. The ejector pump, however, is not usually recommended for wells with depths in excess of 80 ft.

Centrifugal pumps are somewhat critical as to speed and should be used only where power can be applied at a reasonable constant speed. Vertical-type centrifugal pumps are used in deep wells. They are usually driven through shafting by vertical motors mounted at the top of the well. Rather large wells are required for either centrifugal or turbine deep-well pumps, the size depending on the capacity and design of the pump. Centrifugal pumps are efficient in higher capacities, but in the lower capacities of 10 gals. or less per minute, their efficiency is not as high as that of the plunger pumps. It is usually not practical to adopt centrifugal pumps for installations requiring small volumes of water.

Turbine pumps as used in domestic water systems are self-priming. Their smooth operation makes them suitable for applications where noise and vibration must be kept at a minimum.

Ejector pumps are very popular. They operate quietly, and neither the deep-well nor the shallow-well type needs to be mounted over the well.

CHAPTER 9

Sewage Disposal

Municipal Sewage Treatment

Sewage treatment is much more aptly named *waste water treatment*. Waste water is actually 99.94 water by weight. Material suspended or dissolved in the water makes up the remaining .06 percent. Although sewage contains human wastes, it also contains all the waste matter and water discharged from homes, factories, processing plants of all kinds, gasoline service stations, garages, or any other industry discharging wastes into the sewer system. There are three types of sewers:

1. Sanitary sewers carry liquid and water-borne wastes from homes, factories, and industrial and commercial facilities.
2. Storm sewers carry runoff water from streets, roofs, or other drainage not including human wastes.
3. Combination sewers are combined sanitary and storm sewers.

Waste water, after undergoing treatment, will eventually find its way into a river, lake, or other body of water (and in recent years its potential dangers have been recognized). The components of waste water that will deplete the oxygen supply of the stream, lake, or other body of water into which the treated waste water is discharged must be rendered harmless before the waste water effluent is discharged. Certain components of waste water effluent stimulate undesirable growth of plants or organisms and have an undesirable effect on downstream users of the water into which the effluent is discharged. Water contaminated with organic and inorganic materials is technically *polluted*.

The goal of a waste water treatment plant is to simulate as closely as possible the process by which nature cleans and purifies water. Contaminants that have been added to the water, such as phosphates, nitrates, other chemicals, and disease-carrying germs must also be removed or neutralized. Human wastes and other organic matter are consumed or digested by bacteria and other small organisms in water. The bacteria normally present in waste water must have oxygen to consume the raw sewage or other organic matter. A certain amount of dissolved oxygen (DO) is required by a body of water in order to stay alive. When an excessive amount of sewage is dumped into a body of water, the bacteria may use up all the available oxygen in the water in the digestive process. Without oxygen, fish and beneficial plant life die and the water becomes odorous. If the effluent or final discharge from a sewage treatment plant contains organisms that require a large amount of oxygen to aid further bacterial processing, the oxygen supply of the receiving body of water will be greatly depreciated. The measure of the amount of oxygen consumed in the biological process of breaking down organic matter in water is called BOD or *biological oxygen demand*. Therefore, the greater the degree of pollution, the higher becomes the BOD.

All the organics in waste water are not biologically degradable. Pesticides that cannot be broken down biologically may have adverse long-term effects and may contribute to odor, taste, and color problems in downstream water supplies. The COD (chemical oxygen demand) test is used to measure the quantities of nondegradable organics present in the waste water.

Pathogenic bacteria (bacteria that can transmit disease) and viruses are also present in waste water. If suspended solids are present in the final effluent from the treatment plant, the solids can shield the bacteria and virus from contact with disinfecting agents and thus slow down or prevent effective disinfection. Phosphorus and nitrogen, which are also present in waste water, may stimulate the growth of algae in the lakes or streams that receive the final effluent discharge. Algal growths may cause unpleasant taste and odor problems in downstream water supplies. Waste water treatment must be concerned with the removal of phosphorus and nitrogen.

The mineral quality of the potable water furnished by a water utility is changed during water usage. Calcium, sodium, magnesium, chlorides, sulfates, and phosphates are dissolved during water usage and become pollutants. These pollutants are called TDS (total dissolved solids). Control of the TDS present in the final effluent is important because the body of water that receives the final effluent from a treatment plant may be a supply source for a city further downstream.

Primary Waste Water Treatment

Primary waste water treatment is designed to remove both the pollutants that will settle—heavy suspended solids—and those that float, such as grease and oil. Primary treatment will not remove the dissolved pollutants; 60 percent of the raw-sewage suspended solids and 35 percent of the BOD are removed in a typical primary process.

Secondary Treatment

Secondary treatment is designed to remove the soluble BOD left from the primary treatment and to remove more of the suspended solids. Secondary treatment does not remove any appreciable amount of phosphorus, nitrogen, COD, or heavy metals, such as chromium, zinc, lead, silver, cadmium, and mercury. Pathogenic bacteria and viruses are not completely removed in the secondary treatment and may require further treatment before the final effluent is released into the receiving body of water.

Advanced Treatment

Advanced treatment may consist of allowing the final effluent of the secondary treatment plant to flow into irrigation systems serving soil crop acreages. Another method is to chemically treat and filter the final effluent—a process similar to the one used by a utility to process water. It is possible to remove as much as 99 percent of the BOD and phosphorus, all suspended solids and bacteria, and 95 percent of the nitrogen, using some of these processes. The resulting water product is a clean, odorless, colorless, sparkling effluent, indistinguishable from drinking water.

Most of the impurities removed from waste water become "sludge," or suspended solids, in the process. The sludge may be further processed to serve as fertilizer, used as landfill, or incinerated.

Sequence of Waste Water Treatment

Waste water processing or treatment is a combination of processes, both biological and physical, designed to remove organic matter from solution. This is the principle involved in sewage treatment ranging from the individual application, such as a septic tank or other home sewage treatment plant, to the large municipal treatment plants.

Municipalities must handle and treat millions of gallons of liquid and water-carried wastes every day and dispose of the sludge and liquid effluent that are the end products of sewage treatment. The sludge can be treated and used as fertilizer, buried in a landfill, or incinerated. The liquid effluent must be chemically treated to kill any harmful bacteria that may remain after processing and then released into a filter bed, stream, river, or other body of water.

The block diagram shown in Fig. 9-1 illustrates the step-by-step process of treating sewage in one large midwestern municipality.

Sewage entering the treatment plant goes directly to a wet well. A bar screening device in the wet well picks out paper, rags, and other nonsoluble material for transmission to the incinerator. The remaining liquid wastes are lifted by a pump into the raw sewage tower. The raw sewage tower is designed to handle the maximum processing capacity of the treatment plant; any inflow of sewage in excess of the plant handling capacity is bypassed to a nearby creek. The raw sewage passing through the tower goes to the pre-aeration tank and grit chamber. Large blowers located in the blower room in the plant furnish air, which is circulated to jets located under water in the pre-aeration tank. The air, released under water in the tank, causes violent agitation of the water, breaking up the solids

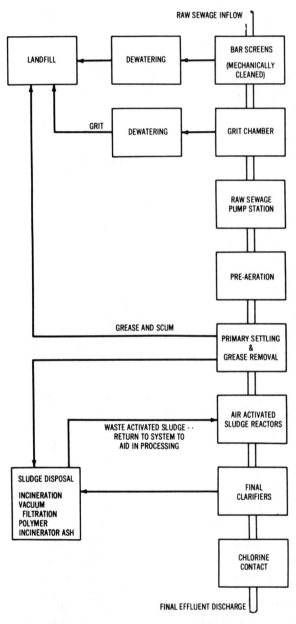

Fig. 9-1. Block diagram of sewage treatment process used by a large midwestern city.

and separating the grit, sand, gravel, and glass from the fecal matter and greases. After a period in the pre-aeration and grit tank, the raw sewage travels to the primary settling tanks.

Primary Clarifiers

Raw waste water enters the circular primary settling tank through ports at the top of a central vertical inlet pipe. The inlet well directs the flow of liquid downward and equally in all directions. As shown in Fig. 9-2 the bottom of the tank is sloped to the center. A collector arm that rotates very slowly pushes the settled solids or sludge to the center of the tank where it is drawn off through the sludge draw-off well. A weir at the outside top edge of the tank permits outflow of liquid wastes only. Floating solids drifting toward the edge of the tank are prevented from discharge by a baffle mounted in front of the weir. A skimmer mounted on the collector arm rotates with the collector arm and collects scum from the surface and deposits it into a scum box for transfer to the incinerator. The liquid effluent flowing over the weir is collected and piped to the air-activated sludge reactors.

Air-Activated Sludge Reactors

Air-activated sludge reactors are concrete tanks equipped with submerged air nozzles or diffusers designed for oxygenation and mixing of the liquid wastes entering the tank. Some of the activated sludge removed from the final settling tanks is recirculated to the activated sludge tank. The sludge, kept in suspension by air, forms contact material; dissolved and very fine organic matter is converted into activated sludge. The liquid waste (containing some sludge) now goes to the final settling tanks for clarification and chlorination.

Final Clarification

The liquid wastes move very slowly through the final settling tanks. Most of the material suspension in these tanks settles out as the liquid travels through. Chlorination of the effluent is the final step in treatment, and after sufficient contact time to insure the satisfactory kill of the remaining organisms, the effluent is permitted to flow into a nearby creek.

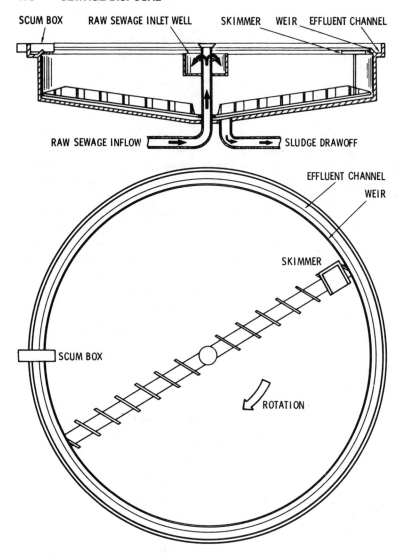

Fig. 9-2.   Circular primary settling tank.

As mentioned, part of the sludge removed from the final settling tanks is recirculated back to the air-activated sludge reactors; the remaining portion is piped to the incinerator site where it will be prepared for incineration.

Preparation for Incineration

When the sludge, which has been removed from the settling tanks, reaches the incinerator site, it is pumped into large circular tanks where it is mixed with ash from the incinerator. This process, called "thickening," is necessary in order for the sludge to be of the proper consistency when it reaches the vacuum filter.

Chemicals such as ferric chloride, lime, or polymers must be added to the thickened sludge to capture the fines which would otherwise be drawn through the filter medium. Polymers are often used because they are easier to feed into the thickened mixture and are often more economical. The polymers are added just before the vacuum filtration process begins.

Vacuum Filtration

The most common mechanical method used to de-water sludge in preparation for incineration is the use of rotary drum vacuum filtration. A cylindrical drum covered with a filter medium slowly rotates while partially submerged in a tank of thickened, chemically treated sludge. The filter medium, whether made of cloth or metal, is porous. As the drum is rotated slowly through the thickened sludge, vacuum is applied under the filter medium. The applied suction draws the sludge to the filter medium and holds it there; at the same time the suction also draws the water from the sludge. The extracted water is collected inside the drum and piped to a drain. As the drum rotates and reaches the release point for the sludge cake, the vacuum is broken and the sludge, now a semi-dry "cake," drops off, ready for final disposal. The filter medium continues to rotate, passing through a spray wash to prepare the filter for sludge pickup as the cycle continues. Vacuum filters are often installed in batteries, with the sludge cake dropping onto an endless belt conveyor and carried directly to the incinerators. A vacuum filter is shown in Fig. 9-3.

Incinerators

Incineration of the sludge cake takes place in a multiple hearth furnace, similar to the one shown in Fig. 9-4. The furnace consists of a circular steel shell with several hearths in a vertical stack and

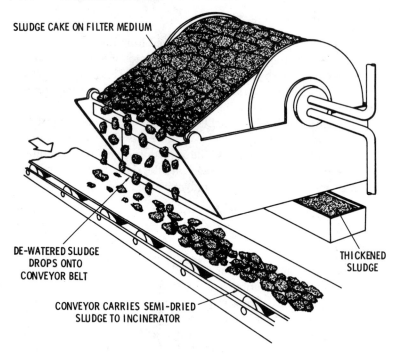

SLUDGE CAKE ON FILTER MEDIUM

DE-WATERED SLUDGE
DROPS ONTO
CONVEYOR BELT

THICKENED
SLUDGE

CONVEYOR CARRIES SEMI-DRIED
SLUDGE TO INCINERATOR

Fig. 9-3.   A vacuum-type sludge filter.

a central rotating shaft with rabble arms. The sludge cake is fed into
the top hearth, the rotating arms spread the cake and push it to the
center where it drops to the second hearth. Here it is again spread
and then pushed to the outside perimeter where it drops to the
third hearth. The two upper-level hearths evaporate any remaining
moisture from the sludge cake and the actual burning of the sludge
starts at the third hearth. Subjected to temperatures between 1400°
and 1500°F, the sludge is incinerated in the third and fourth hearths
and the ash residue is cooled in the lower levels.

The smoke created in the process contains a high percentage
of fly ash. This is removed by passing the smoke through a cyclonic
wet scrubber before it is released to the atmosphere. The ash from
the scrubber, combined with the ash from the incinerator, is mixed
into a slurry and piped to the thickening tanks to be mixed with
the incoming sludge.

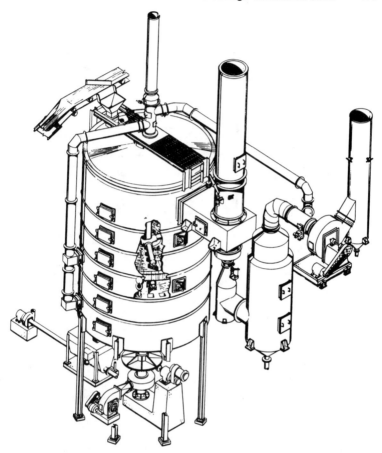

Fig. 9-4.   A multiple hearth incinerator. *(Courtesy Nichols Engineering and Research Corp.)*

Individual Home Sewage Treatment Plants

Individual home-sized sewage treatment plants use the same basic purification process as large central plants. Fig. 9-6 shows a cutaway view of a home-sized plant.

The *primary* treatment compartment receives the household sewage and holds it long enough to allow solid matter to settle to the sludge layer at the bottom of the tank. Organic solids are here

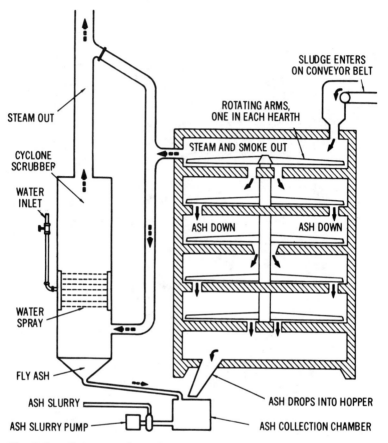

Fig. 9-5. Cut-away view of a multiple hearth incinerator. *(Courtesy Nichols Engineering and Research Corp.)*

broken down physically and biochemically by anaerobic bacteria (bacteria that live and work in the absence of oxygen). Grit and other untreatable materials are settled out and contained in the primary chamber. The partially broken down, finely divided material at the top of the primary compartment is passed on to the aeration chamber.

In the *aeration* chamber the finely divided, pretreated material from the primary compartment is mixed with activated sludge and aerated. Aerobic bacteria (bacteria that live and grow in the presence

Fig. 9-6.   A home-sized sewage treatment plant. *(Courtesy Jet, Inc.)*

of oxygen) further digest the material which enters this chamber. Large quantities of air are injected into this compartment to hasten the digestive process. The aerobic bacteria use the oxygen in solution to break down the sewage and convert it into odorless liquids and gases. After treatment in this chamber, the liquid flows into the settling-clarifying compartment.

The final phase of the operation takes place in the *settling-clarifying* compartment. Here a tube settler eliminates currents and encourages the settling of any remaining settleable material, which is returned, by the sloping end wall in the tank, to the aeration chamber for further treatment. A nonmechanical surface skimmer, operated by hydraulics, skims any floating material from the surface of the settling compartment and returns it to the aeration chamber. The remaining odorless, clarified liquid flows into the final discharge piping through the baffled outlet.

If local health regulations require it, a nonmechanical chlorinator can be added to the plant to provide chemical sterilization of the effluent. If tertiary treatment is required, a separate up-flow filter providing further biological treatment from bacterial growth on the filter medium can be added. A chlorinator may also be added in the tertiary tank if desired. There are many areas where, due to soil composition, a septic tank and disposal field cannot provide a satisfactory solution to the problem of sewage disposal. The effluent from a septic tank requires further treatment by aerobic bacteria, in a finger system or a disposal field. If properly installed and maintained, a home-sized sewage treatment plant, such as is shown in Fig. 9-6, will produce an effluent that either reduces the requirements for subsurface filters (finger systems/disposal fields) or eliminates them entirely.

Private Sewage Disposal Systems

Septic Tanks and Disposal Fields

A septic tank is a water-tight receptacle which receives the discharge of a drainage system and is designed and constructed so as to separate solids from liquids, digest organic material through a period of detention, and allow the liquids to discharge through a system of open-joint or perforated piping into a disposal field or disposal pit.

Structures in areas not served by a public sewer may install septic tanks to dispose of sewage provided certain criteria established by the Administrative Authority are complied with:

1. *When no public sewer, intended to serve any lot or premises, is available in any thoroughfare or right-of-way abutting such lot or premises, drainage piping from any structure shall be connected to an approved private sewage disposal system.*

2. *The type of system shall be determined on the basis of location, soil porosity, ground water level, and shall be designed to receive all sewage from the property.*

3. *No private sewage disposal system shall be located in any lot other than the lot which is the site of the structure served by the private sewage disposal system.*

4. *The proposed site shall be subjected to percolation tests acceptable to the Administrative Authority.*

5. *Septic tank design shall be such as to produce a clarified*

*effluent consistent with accepted standards and shall provide
adequate space for sludge and scum accumulations.*

6. *Septic tanks shall be constructed of solid durable materials,
   not subject to excessive corrosion or decay, and shall be
   water-tight.*

7. *Access to each septic tank shall be provided by at least two
   (2) manholes twenty (20) in. in dimension or by an equivalent
   removable cover slab.*

8. *Each septic tank shall be structurally designed to withstand
   all anticipated earth loads.*

9. *Septic tanks installed under asphalt or concrete paving shall
   have the manholes accessible by extending the manhole
   openings to grade level.*

Note: The above quotations from the UNIFORM PLUMBING CODE
are presented to show the reader that installations of septic tanks
are strictly regulated. The quotations represent only a very small
part of the total regulations governing septic tank installations and
are not to be used for any specific installation. The Code adopted
for use in an area where an installation is planned must govern the
installation.

All private sewage disposal systems shall be so designed that
additional subsurface drain fields and/or seepage pits, equivalent to
at least 100 percent of the required original system, may be installed
if the original system cannot absorb all the effluent discharged from
the system.

The liquid capacity of a septic tank is determined by the number
of bedrooms, the estimated waste or sewage flow, or the number
of plumbing fixture units in structures which the septic tank serves.
The capacity of any one septic tank and its drainage system is limited
by the soil structure classification of the drainage field or seepage
pit.

Bacterial action is the basis of all septic tank systems. There
are two types of bacteria at work in a properly operating septic tank
and disposal field or seepage pit. One type is *anaerobic* bacteria,
bacteria which live and work in the absence of oxygen. Soil and
waste products enter the primary chamber of the septic tank and
are attacked by anaerobic bacteria. Solids are broken down and
liquefied, and gases form, carrying lighter materials to the top forming a scum in which the bacteria multiply. Particles which cannot

be digested sink to the bottom and form sludge. Most of the bacterial action in a septic tank takes place in the inlet or primary compartment.

Septic tanks must have a minimum of two compartments. The primary compartment must be not less than two-thirds of the total capacity of the tank. The compartments are separated by baffles and as the digesting process continues, clarified liquids are carried into the outlet compartment. Each time wastes enter the tank an equal volume of liquid is displaced and discharged through the outlet piping of the tank. The outlet piping is designed to permit liquids only to be discharged. Although relatively clear liquids leave the tank, the digestive process is not yet completed. The liquids leaving the tank flow through a distribution box and into a drainage field, a seepage pit, or both. Bacteria called *aerobic* bacteria, which thrive in the presence of oxygen, now go into action and further digest the liquid effluent. The size of the drainage field and/or seepage pit is of vital importance to the action of a septic tank system; the requirements of an installation must be decided on an individual basis. The basic design of a septic tank is shown in Fig. 9-7.

Commercial Waste Water Treatment Plants

Housing sub-divisions, motels, hotels, schools, office buildings, commercial/industrial complexes and shopping centers are often built in outlying areas not served by public sewers. The sewage and waste water generated by these establishments must be treated and disposed of in a way that will not harm the environment.

There are two ways to solve this problem:

1. Construction of an on-site or nearby sewage treatment plant.
2. Installation of one or more "package" sewage treatment plants.

Construction of an on-site plant which will use the same methods to treat waste water as the large municipal plants use may take months and may create problems at a future date if expansion of the existing plant is needed due to growth of the input facilities.

Package waste water treatment plants are built in sizes to accommodate the flow load, expressed in terms of gallons per day, the BOD (biological oxygen demand) per day, or both. If the input facilities grow, additional units may be added to handle the increased load.

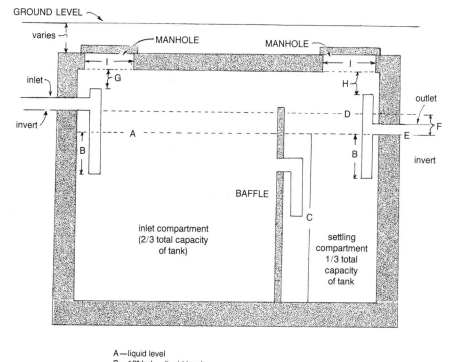

GROUND LEVEL

varies

inlet

invert

MANHOLE

MANHOLE

inlet compartment
(2/3 total capacity
of tank)

BAFFLE

settling
compartment
1/3 total
capacity
of tank

outlet

invert

A—liquid level
B—12″ below liquid level
C—midway in liquid depth
D—invert of inlet
E—invert of outlet
F—minimum distance below inlet invert
G—minimum 2″
H—minimum 4″
I —minimum 20″

Fig. 9-7.   Basic construction of a septic tank.

Package treatment plants are designed to treat sewage using the same methods employed by large municipal plants: extended aeration and aerobic digestion. The process is basically a three-stage process:

1. Pre-treatment.
2. Aeration.
3. Settling.

In the pre-treatment stage, large objects are caught by bar screens and trash traps. A comminutor (waste water grinder) reduces sewage to a semi-liquid state.

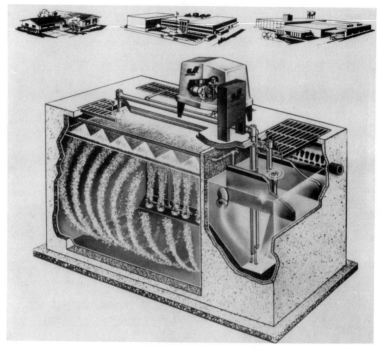

Fig. 9-8. A commercial waste water treatment plant. *(Courtesy Jet, Inc.)*

After pre-treatment, the waste water flows into an aeration tank where it is mixed with air. Air diffusers in the bottom of the aeration tank bubble in large amounts of air for two purposes: (a) to meet the oxygen demand of the aerobic digestion process and (b) to mix the aeration tank contents, insuring complete treatment. In the aeration process the pre-treated waste water is transformed into a clear odorless liquid. From the aeration tank the liquid flows into a settling tank that holds the liquid completely still. Any small particles in suspension settle to the bottom and are returned to the aeration tank for further treatment. The settling process in the final tank of the plant shown in Fig. 9-8 leaves a clear highly treated effluent ready for return to the environment. The final effluent must be disposed of, and this can be accomplished in several ways:

1. Allowed to run into a nearby stream or body of water.
2. Absorbed into the ground through a dispersal field.
3. Absorbed into the ground through a sand filter bed.

Fig. 9-9. A non-mechanical chlorine dispensing system. *(Courtesy Jet, Inc.)*

Local regulations may require the chlorination of the final effluent before disposing of the effluent. Proper chlorination insures positive disinfection to kill bacteria and inhibit bacteria re-growth. A non-mechanical chlorine dispensing system is shown in Fig. 9-9.

CHAPTER 10

Questions and Answers for Plumbers and Pipe Fitters

As mentioned earlier in this book, plumbers are called on to know a wide variety of information. What follows is a series of questions designed to evaluate the depth and degree of an individual's knowledge. The answers, it should be noted, are based on common practice. Individual communities or jurisdictions may have different requirements, and if so, these must be abided by.

What is meant by plumbing?
Answer: It is the art of installing the pipes, fixtures, and other apparatus in buildings for bringing in the water supply and removing liquid and water-carried wastes.

What constitutes the plumbing system of a building?
Answer: It includes the water-supply distributing pipes, the fixtures and fixture traps, the soil, waste, and vent pipes, the building drain and building sewer, the storm-water drainage, all with their devices, appurtenances, and connections within and adjacent to the building.

What is a water service pipe?
Answer: It is the pipe from the water main to the building served.

What is meant by water distribution pipes?

Answer: They are the pipes that carry water from the service pipes to the plumbing fixtures.

What are plumbing fixtures?

Answer: These are receptacles intended to receive and discharge water, liquid, or water-carried wastes into a drainage system.

Why must all plumbing fixtures be connected to a water source?

Answer: In order that they be provided with a sufficient amount of water to keep them in a serviceable and sanitary condition.

What is a trap?

Answer: A fitting or device constructed to prevent the passage of air or gas through a pipe without materially affecting the flow of sewage or waste water through it.

What is meant by the vent pipes?

Answer: Any pipe provided to ventilate a house drainage system and to prevent trap siphonage and back pressure.

What is a local ventilating pipe?

Answer: A pipe through which foul air is removed from a room or a fixture.

What is a soil pipe?

Answer: Any pipe which carries the discharge of water closets, with or without the discharges from other fixtures, to the building drain.

What is a waste pipe?

Answer: A pipe that receives the discharge of any fixture except water closets, and carries it to the building drain, soil, or waste stacks.

What is a main?

Answer: The main of any system of horizontal, vertical, or continuous piping is the part of the system that receives the wastes and vents or back vents from fixture outlets or traps directly or through branch pipes.

What is meant by branch piping?

Answer: The part of the system that extends horizontally at a slight grade, with or without lateral or vertical extensions or vertical arms, from the main to receive fixture outlets not directly connected to the main.

What is a stack?

Answer: Stack is a general term for any vertical line of soil, waste, or vent piping.

What is meant by a building drain?

Answer: The part of the lowest horizontal piping of a building drainage system that receives the discharge from soil, waste, and other drainage pipes inside the walls of any building and carries it to the building sewer beginning 5 ft. outside of the building wall.

What is a building sewer?

Answer: It is that part of the horizontal piping of a building drainage system extending from the building drain 5 ft. outside of the inner face of the building wall to its connection with the main sewer and conveying the drainage from one building site.

What is a dead end?

Answer: A branch leading from a soil, waste, vent, building drain, or building sewer, which is terminated at a developed distance of 2 ft. or more by means of a cap, plug, or other fitting not used for admitting water to the pipe.

What are the rules as to workmanship and materials?

Answer: All work must be performed in a thorough, workmanlike manner, and all material used in any drainage or plumbing system or part thereof must be free from defects.

What is a combination waste and vent system?

Answer: A combination waste and vent system is a specially designed system of waste piping embodying the horizontal wet venting of one or more sinks or floor drains by means of a common waste and vent pipe adequately sized to provide free movement of air above the flow line of the drain.

What kind of material must be used in main, soil, waste, or vent pipes?

Answer: Main, soil, waste, and vent pipes can be made of whatever local codes allow, usually cast iron, steel, copper, or plastic.

What is meant by the term riser lines in a plumbing system?

Answer: The term *riser* is generally applied to the vertical pipes extending through the building from its connection with the house main.

What are the requirements when brass and copper pipe are used?

Answer: All brass and copper pipe shall conform to the standard specifications of the A.S.T.M. (American Society for Testing Materials).

How should a caulked joint be made?

Answer: All caulked joints should be firmly packed with oakum or hemp, and secured only with pure lead not less than 1 in. deep, well caulked. No paint, varnish, or putty should be used until the joint is tested.

How is an approved joint in lead pipe or between lead and brass or copper pipe made?

Answer: Joints in lead pipe, or between lead pipe and brass or copper pipe, ferrules, soldering nipples, bushings, or traps, in all cases on the sewer side of the trap, and in concealed joints on the inlet side of the trap, should be full-wiped joints with an exposed surface of the solder to each side of the joint of not less than $\frac{3}{4}$ in., and a minimum thickness at the thickest part of the joint of not less than $\frac{3}{8}$ in.; or by use of a lead-to-iron union.

What are the requirements when joints are made between lead and cast or wrought iron?

Answer: The joints shall be made by means of a caulking ferrule, soldering nipple, or bushings.

When should slip joints and unions be used?

Answer: Slip joints should be permitted only in trap seals or on the inlet side of the trap. Unions on the sewer side of the trap should be ground-faced and should not be concealed or enclosed.

What is the name used for a fitting that has one side outlet at right angles to the run?

Answer: Tee.

What is the name for a fitting having a larger size at one end than on the other?

Answer: Reducer.

What is the name of a fitting which has one side outlet at any angle other than 90°?

Answer: Wye.

What is the term generally employed for a piece of pipe threaded on both ends and not more than 12 in. long?

Answer: Nipple.

What should be the minimum distance between the hot- and cold-water risers in a plumbing system?

Answer: The distance between the hot- and cold-water risers where a hot-water supply is installed should not be less than 6 in., and where conditions encountered are such that they cannot be readily placed 6 in. or more apart, the hot-water riser should be covered with an approved insulating material so it doesn't interfere with the prompt delivery of hot water to the faucet when required.

What is the minimum size of a main vent pipe?

Answer: The size of a main vent pipe must never be less than 3 in. in diameter.

How are the required sizes of vent pipes determined?

Answer: The diameter of a stack vent, vent stack, or relief vent shall not be less than one-half of the diameter of the drain served, but in no case less than $1\frac{1}{4}$ in. and shall be determined from its length and the total fixture units connected thereto. The size and length of vent piping must be as specified in the plumbing code adopted in the area of installation.

Where should a clean-out be placed in a vertical waste or soil stack?

Answer: A clean-out that is easily accessible should be installed at the base of each vertical waste or soil stack.

Should each buiding have soil and waste stacks?

Answer: Yes. Every building in which plumbing fixtures are installed shall have a soil and waste stack extending full size through the roof.

How far above the roof should soil or vent pipe lines be carried?

Answer: All roof extensions of soil and vent stacks shall be run full size at least 6 in. above the roof coping, and when the roof is used for other purposes than weather protection, such extension shall not be less than 5 ft. above the roof.

Should special rules apply to soil and vent pipes used in a cold climate?

Answer: Where there is danger of frost closing it, no roof extension shall be less than 4 in. in diameter. The change in diameter must be accomplished by the use of a long increaser at least 1 ft. below the roof; and where the access to the roof is difficult, a test opening should be provided at this point.

May a vent or soil pipe be terminated within a distance of 2 ft. from any door, window, scuttle, or air shaft?

Answer: No. The roof terminal of any stack or vent, if within 10 ft. of any door, window, scuttle, or air shaft, shall extend at least 3 ft. above the same.

May soil or vent lines be carried outside of buildings?

Answer: No soil or vent lines shall be installed or permitted outside of a building unless adequate provision is made to protect them from frost.

Where shall main vents be connected?

Answer: All main vents or vent stacks shall connect full size at their base to the main soil or vent pipe at or below the lowest fixture branch, and shall extend undiminished in size above the roof, and shall be reconnected with main soil or waste vent at least 3 ft. above the highest fixture branch.

When offsets are made in vent lines, how should they be connected?

Answer: All vent and branch vent lines shall be connected and installed so that they are free from drops or sags, and be graded and connected to drip back to the soil or waste pipe by gravity. Where the vent pipes connect to a horizontal soil or waste pipe, the vent branch shall be taken off above the center line of the pipe, and the vent pipes must rise vertically, or at an angle of 45°, to a point 6 in. above the fixture it is venting before offsetting horizontally or connecting to the branch, main waste, or soil vent.

When may circuit or loop vents be employed?

Answer: A circuit or loop vent is permitted as follows: A branch soil or waste pipe to which two and not more than eight water closets, pedestal urinals, trap standards, slop sinks, or shower stalls are connected in series may be vented by a circuit or loop vent, which shall be taken off in front of the last fixture connection. Where fixtures discharge above such branches, each branch shall be provided with a relief one-half the diameter of the soil or waste stack, taken off in front of the first fixture connection.

What is the required running diameter of traps for urinals?
Answer: $1\frac{1}{2}$ in. (minimum).

What are the requirements for a permissible trap?

Answer: Every trap shall be self-cleaning. All traps used for bathtubs, lavatories, sinks, and other similar fixtures should be of brass, cast iron, galvanized malleable iron, or porcelain-enameled inside. Galvanized or porcelain-enameled traps should be extra heavy and have a full-bore smooth-interior waterway, with threads tapped out of solid metal. Some local codes allow the use of plastic materials and traps.

Where should the fixture trap be placed relative to its fixture?

Answer: The trap should be as close to the fixture as possible, but not more than 24 in. from the fixture.

What are the requirements in regard to clean-outs in fixture traps?

Answer: All traps, except water-closet traps, shall be provided with an accessible brass trap screw of ample size, protected by the water seal.

May fixture traps be connected in series?

Answer: No. No fixture shall be double-trapped.

Must all fixture traps be protected against back pressure and siphonage?

Answer: Yes. Every fixture trap shall be protected against siphonage and back pressure, and air circulation assured by means of a soil or waste stack vent, a continuous waste or soil vent, or a loop or circuit vent. Crown vents are not permitted.

Must trap levels be protected against frost and evaporation?

Answer: All traps shall be installed true with respect to their water seals and protected from frost and evaporation.

What is the seal of a trap?

Answer: It is the depth of the water between the dip and the outlet of the trap. The effectiveness of a trap always depends on its water seal.

What is the dip of a trap?

Answer: The part of a trap that dips into the seal, and under which all waste matter must pass.

What is meant by the term siphonage?

Answer: By referring to an "S" trap, it will be readily seen that the outlet forms a perfect siphon, the part of the trap between the dip and the outlet forming the short side, and the waste pipe from the outlet downward forming the long side. When a large quantity of water is discharged from the fixture into the trap, the water fills the entire trap and waste pipe for some distance below the trap. It can be seen that the weight of the water is much greater at the outlet side than at the inlet side of the trap, and it tends to cause the water in the trap to rise to the outlet and follow the larger body of water in the waste pipe, leaving the trap without any water to form its water seal.

What determines the resistance against siphonage in a trap?

Answer: The depth of the water seal determines the amount of resistance a trap will offer to being unsealed by siphonage.

How should service pipes be protected when exposed to frost?

Answer: They should be protected by a sufficient amount of felt or other insulation and supported by metal sleeves or approved metal bands.

What precautions should be taken when thawing a frozen water pipe?

Answer: If the thawing is done with a blow torch or hot water, the thawing medium should be applied to the water-supply end of the pipe, opening a faucet if possible to indicate when the flow of water starts. It is well to keep in mind that the middle of the pipe should never be thawed first, because expansion of the water confined by ice on both sides may burst the pipe.

How should a waste or sewer pipe be thawed out?

Answer: When thawing a waste or sewer pipe, always work upward from the lower end, to permit the water to drain away.

Briefly discuss the most effective methods used to thaw out frozen water pipes.

Answer: The method to be used will be determined by the amount of pipe to be thawed, as well as the size and location. For short lengths of exposed pipe, boiling hot water or hot cloths have proven to be effective. If there is no danger of fire, or if the necessary precautions against fire are observed, a blow torch or burning newspaper run back and forth along the frozen water pipe gives quick results. When the pipe to be thawed is located underground, or is otherwise inaccessible, the pipe should be disconnected at the house end and boiling water poured through the opening, using a small piece of auxiliary pipe or rubber tubing to which a funnel is attached conveniently. When a long section of inaccessible piping or leaders are to be thawed, low-voltage electricity has been found to be effective, particularly with electric heating cables. Electric blankets have been used with success as well.

What are the causes of sewer obstruction, and how should sewers be constructed to lessen this trouble?

Answer: Causes may be any one of the following: broken pipes, insufficient grade to give cleansing velocities, newspaper, rags, garbage or other solids in the sewage, congealing of grease in pipes and house sewer traps, and poor joint construction whereby rootlets grow into the sewer and choke it. The proper grade and good construction, with particular care given to the joints, will avert or lessen these troubles. The sewer should be made perfectly straight with the interior of the joints scraped or swabbed smooth. When the joint-filling material has set, the hollows beneath the hubs should be filled with good earth free of stones, well tamped or puddled in place. It is important that like material be used at the sides of the pipe and above it for at least one foot.

What is a siphon chamber and how does it work?

Answer: The purpose of a siphon chamber is to secure intermittent discharge, thus allowing a considerable period of time for one dose to work off in the soil and for air to enter the soil spaces before another flush is received. It is also used to secure distribution over a larger area and in a more even manner than where the sewage is allowed to dribble and produce the conditions of the old-fashioned sink drain, namely, a small area of water-logged ground.

Three types of sewage siphons are shown in Fig. 10-1. In all, the essential principle is the same. A column of air is entrapped between two columns of water; when the water in the chamber rises to a predetermined height, called the discharge line, the pressure forces out the confined air, upsetting the balance and causing a rush of water through to the sewer. The entire operation is fully automatic and very simple. The siphons shown are commercial products and are made of cast iron. Their simplicity and reliability are enhanced by the small number of nonmovable parts.

Manufacturers furnish information for setting the siphons and putting them in operation. For example, for type 2 in Fig. 10-1: (1) set siphon trap (U-shaped pipe) plumb, making E (height from floor to top of long leg) as specified; (2) fill siphon trap with water till it begins to run out at B; (3) place bell in position at top of long leg, and the siphon is ready for service. Do not fill vent pipe on side of bell. The overhead siphon (type 3, Fig. 10-1) may be installed readily in a tank already built by the addition of an outlet pump. If properly set and handled, sewage siphons require very little attention and flush with certainty. However, like all plumbing fixtures, they are subject to stoppage if rags, newspapers, and similar solids

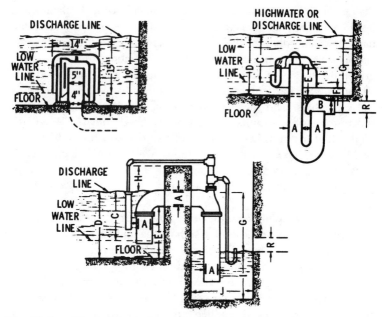

Fig. 10-1. Three types of sewer siphon systems.

get into the sewage. If fouling of the sniffling hole or vent prevents the entrance of sufficient air into the bell to lock the siphon properly, allowing sewage to dribble through, the remedy is to clean the siphon. It is well to remember that siphons are for handling only liquid; sludge, if allowed to accumulate, will choke them.

Is it necessary to have a plumbing system tested and inspected after completion, and who should make the test?

Answer: The entire plumbing system shall be tested by the plumber in the presence of a plumbing inspector, or the proper administrative authority, to insure compliance with all the requirements of the plumbing regulations, and to insure that the installation and construction of the system is in accordance with the approved plans and the permit.

How is this test accomplished?

Answer: By filling all the piping of the plumbing system with water or air. After the plumbing fixtures have been set and their traps filled with water, the entire drainage system shall be submitted to the final air-pressure test. The proper administrative authority may require the removal of any clean-outs to ascertain if the pressure has reached all parts of the system.

How should the water test be made?

Answer: The water test may be applied to the drainage system in its entirety or in sections. If applied to the entire system, all openings in the piping shall be tightly closed, except the highest opening above the roof, and the system filled with water to the point of overflow above the roof. If the system is tested in sections, each opening shall be tightly plugged, except the highest opening of the section under test, and each section filled with water; but no section shall be tested with less than a 10-ft. head of water or with less than 5 lbs. of air pressure. In testing successive sections, at least the upper 10 ft. of the next preceding section shall be retested, so that no joint or pipe in the building is tested for less than a 10-ft. head of water or 5 lbs. of air pressure.

How should the air test be made?

Answer: By attaching the air compressor or test apparatus to any suitable opening, and closing all other air inlets and outlets to the system, then forcing air into the system until there is uniform pressure of 5 lbs. per sq. in. (psi), or sufficient to balance a 10-in. column of mercury.

How long shall this air pressure be maintained in the drainage system?

Answer: For at least 15 minutes.

How should the final air test be made?

Answer: In the final air test, the air machine shall be connected to any suitable opening or outlet, and air pressure equivalent to a 1-in. water column shall be applied and left standing at least 15 minutes. If there is no leakage or forcing of trap seals indicated by the fluctuation of the drum, float, or water column, the system is air-tight.

In what order may the tests be made?

Answer: Separately, or as follows: (1) The building sewer and all its branches from the property line to the building drain. (2) The drain and yard drains, including all piping to the height of 10 ft. above the highest point on the house drain, except the exposed connections to fixtures. (3) The soil, waste, vent, inside conductor, and drainage pipe which should be covered up before the building is enclosed or ready for completion. [The tests required for (2) and (3) may be combined.] (4) The final test of the whole system. (5) After each of the tests has been made, the proper administrative authority shall issue a written approval.

What is a relief valve?

Answer: It is a valve arranged to provide an automatic relief in case of excess pressure.

What is a safety valve?

Answer: It is a relief valve for expansive fluids and is provided with a chamber to control the amount of blow-back before the valve reseats.

What is a stop valve?

Answer: It is a valve of the globe type used to shut off a line.

What is a back-pressure valve?

Answer: It is a valve similar to a low-pressure safety valve that is set to maintain a certain back pressure on feed operating pressure irrespective of pressure variations of the supply. The back-pressure valve is arranged to relieve any excess supply to the atmosphere or elsewhere, and it opens and closes automatically as required to produce this result.

What is meant by the term electrolysis?

Answer: It is generally applied to electrolytic corrosion due to electric current conduction by water, gas mains, or metallic structures.

Where does electrolysis take place?

Answer: Along water mains or metallic structures, where the electric stray current leaves the metal for the ground or some other conductor of less resistance.

Where is the electrolytic corrosion most common?

Answer: In densely populated areas along electric railroad lines where track rails are utilized as a negative return circuit.

How can electrolysis be avoided or lessened?

Answer: By lowering the voltage drop, which is done by increasing the metallic area of the negative return circuits adjacent to the water mains. In some cases, insulating or installing high-resistance pipe joints has been found to limit the conduction of stray electrical currents.

How may an electric current be detected in a main?

Answer: By means of a sensitive galvanometer, which can be calibrated to show the potential drop along the pipe, measuring the distance, and calculating the cross-sectional area. The potential drop divided by the resistance gives the flow of current in the main. By knowing the direction of the flow, its amount, and the efficiency of corrosion, the actual damage being done by electrolysis may be calculated as a definite weight of metal per annum.

What are the essential requirements of piping and apparatus for fire protection?

Answer: It must be capable of producing, without question, the desired performance, and it must be designed to function invariably, regardless of age or weather conditions.

Who establishes rules for all kinds of fire protective apparatus?

Answer: The National Board of Fire Underwriters and allied organizations.

How should the piping for an automatic sprinkler system be designed?

Answer: It must be designed so as to insure: (1) an adequate and reliable water supply, (2) ample and complete distribution, (3) proper protection against freezing.

What are the rules in regard to sprinkler system water supply?

Answer: It is generally considered necessary to have two sources of water supply, one which should require no manual operation.

For example, a common arrangement is a gravity tank in combination with a fire-department connection to be used when the apparatus arrives.

What are the N.B.F.U. rules with regard to location and spacing of sprinkler heads?

Answer: They take into account the type of building construction and the dimension of the bays. In general, one sprinkler head is required for each 80 to 100 sq. ft. of floor area.

What is the relation between the number of heads and branch pipes?

Answer: Piping should be arranged that the number of heads on any branch pipe does not exceed eight.

What should be the size of riser lines in a sprinkler system?

Answer: Each riser in a sprinkler system should be of sufficient size to supply all the sprinklers connected to it on any one floor, or, if there is no approved fire stop between the floors, the riser should be of sufficient size to accommodate the total number of sprinklers.

Must riser and supply lines be protected against frost?

Answer: If in exposed locations, they must be adequately protected against frost by means of insulating materials.

What type of valves may be used on a fire-protective system?

Answer: All valves must be of the OS & Y pattern, and check valves should be installed in all sources of supply. Each system should be provided with a gate valve located to control all sources of water supply except that from fire-departmental sources.

What is a *dry system,* and where is such a system required?

Answer: It is a system in which the piping is ordinarily filled with air at a pressure considerably lower than water pressure. When a sprinkler head opens, water enters the system and drives the air out ahead of it. This type of system is required in rooms that cannot be properly heated.

What is the most important feature of the dry system?

Answer: It is the dry-pipe valve, a device that normally prevents water from entering the system but that opens when the air pressure is lowered due to the opening of a head.

What is the water source for an automatic sprinkler system?
Answer: A municipal system under pressure or overhead gravity tanks are used for automatic sprinklers and hose connections.

How are the required tank sizes determined?
Answer: The size of a tank for a given service is determined individually by the insurance authorities. In general, when feeding sprinklers, the tank must have a capacity of 10,000 to 25,000 gals. When feeding both sprinklers and hose, a minimum capacity of 30,000 gals. is usually required.

What is the size of discharge pipes relative to tank capacity in an elevated gravity-tank system?
Answer: Elevated gravity tanks must have a discharge pipe of not less than 6 in. for tank sizes up to 25,000 gals. capacity, and generally not less than 8 in. for 30,000 up to 110,000 gals., and 10 in. for greater capacities.

How are the tanks protected against freezing?
Answer: The usual arrangement consists of a tubular steam heater to which a connection is made from the base of the tank discharge pipe. The heated water is carried up to the tank by a separate pipe. This arrangement permits the temperature of the coldest water to be observed readily and is by far the simplest and most reliable method. The coldest water should not be allowed to be below 40°F.

What are the regulations in regard to pressure tanks?
Answer: Pressure tanks for fire service are ordinarily kept two-thirds full of water, and with an air pressure on the surface of the water of 75 lbs. plus three times the pressure caused by the column of water in the sprinkler system above the tank bottom.

What is the capacity of pressure tanks?
Answer: The capacity is usually set by the insurance inspection authorities having jurisdiction, and is usually between 4500 and 9000 gals. per tank.

How should a pressure tank be designed?

Answer: It should be in accordance with the rules for unfired pressure vessels of the A.S.M.E.

What determines the use of house supply tanks?

Answer: When the water pressure is not sufficient to supply all fixtures freely and continuously, a house supply tank should be provided. The tank should be adequate to supply all fixtures at all times.

What methods should be used to supply house tanks?

Answer: Tanks should be supplied from the street pressure or, when necessary, by power pumps. When such tanks are supplied from the street pressure, ball cocks should be provided.

Where should the water supply inlet to roof tanks be located?

Answer: Water supply inlets to roof tanks should be located at least 2 in. above the overflow pipe level of the tank and should be equipped with an automatic ball stop. The outlet from a roof tank to the distribution system in the building should be effectively equipped to prevent solids from entering into such piping. All down-feed supplies from a tank, cross-connected in any manner with distribution supply piping in a building supplied by direct street main or pump pressure, should be equipped with a check valve to prevent backflow of water into the roof tank.

May a gravity tank be directly connected to the city water main?

Answer: No gravity tank shall be directly connected to the city water main, but shall be provided with an over-the-rim filler, the orifice of the outlet of which must be elevated a distance equal to the least diameter of such water-discharging orifice or outlet, and in no case less than 1 in. above the top rim of the tank.

What size discharge pipe should be provided for a gravity tank having a capacity of 500 gals. or more?

Answer: The discharge pipe from a gravity tank of 500 gals. or more capacity shall be at least 4 in. nominal diameter for a distance of not less than 4 ft., and in no case should it be smaller than the main section of the riser. The shutoff valve should be the same size as the outlet from the tank, but not less than a 4-in. gate valve.

What are the rules as to the tightness of plumbing joints and connections?

Answer: Joints and connections shall be made gas- and water-tight.

What type of joint is required in vitrified clay pipes?

Answer: Joints in vitrified clay sewer pipe should be firmly packed with oakum or hemp and shall be secured with cement, mortar, or asphaltic compound at least 1 in. thick.

How should caulked joints be made?

Answer: Brass caulking ferrules shall be either of the best quality of cast brass or shall be of the cold-drawn seamless tube variety. Soldering nipples shall be of brass pipe (iron-pipe size) or heavy cast brass of at least the weight shown in Table 10-1. Soldering bushings shall be of brass pipe (iron-pipe size) or heavy brass or copper.

What type of screw joints shall be made in a plumbing system?

Answer: Screw joints shall be tapered with the threads sharp and true, and all burrs due to cutting reamed out smooth. Where fitting compounds, red lead, white lead, or other joint materials are used in making up threaded joints, such materials shall be applied to the male threads only.

What type of wiped solder joints shall be made in a plumbing system?

Answer: Joints in lead pipes, brass or copper pipes, ferrules, soldering nipples, bushings, or traps should be full-wiped joints, either manufactured or made in the field. An exposed surface of the solder of at least $\frac{3}{4}$ in. with a minimum thickness at the thickest part of the joint of $\frac{3}{8}$ in. shall be on each side of the joint. It is unlawful to use overcast or cup joints.

Table 10-1. Weight of Soldering Nipples

Pipe Size (inches)	Actual inside diameter (inches)	Length (inches)	Weight	
			Pounds	Ounces
2.......	2¼	4½	1	. . .
3.......	3¼	4½	1	12
4.......	4¼	4½	2	8

What are the rules for making joints of lead to cast iron, steel, or wrought iron?

Answer: Joints of lead to cast iron, steel, or wrought iron shall be made by means of a caulking ferrule, soldering nipple, or bushing.

What type of fixture flanges shall be used in a plumbing system?

Answer: Flanges to receive fixture outlets should be at least ³/₁₆ in. thick and shall be made of brass or cast iron. (Some plumbing codes allow the use of plastic pipe.)

How are water closets and pedestal urinals to be connected to soil and waste piping?

Answer: Water closets and pedestal urinals shall be connected by means of flanges caulked to cast-iron soil pipe, flanges wiped or soldered to lead pipe, flanges soldered to copper pipe, or, if plastic pipe is permitted, PVC flanges cemented to plastic pipe.

When are slip joints and unions permitted?

Answer: Slip joints or unions are permitted only in trap seals or on the inlet side of the trap, except where it is impracticable to otherwise provide for expansion in stacks of unusual height. The authorities may permit the use of an approved type of expansion joint that comprises, in part, a slip joint.

What are the rules as to roof joint connections in a plumbing system?

Answer: Where the pipes pass through roofs, the joints shall be made water-tight.

When are expansion and contraction of piping due to temperature variations provided for in a plumbing system?

Answer: In structures over 150 ft. high, adequate means shall be provided for taking care of the expansion and contraction of all vertical lines of pipe.

What is the rule for protection of soil or waste stacks?

Answer: Soil or waste stacks shall be installed inside the structure.

What plumbing connections are prohibited?

Answer: It is unlawful to make any waste connection to a bend of a water closet or similar fixture. It is unlawful to use soil or waste vents as soil or waste pipes.

What are the rules for changes in direction in a plumbing system?

Answer: Changes in direction shall be made by the appropriate use of 45° wyes, half wyes, long-sweep $\frac{1}{4}$ bends, $\frac{1}{6}$, $\frac{1}{8}$, or $\frac{1}{16}$ bends, or long-turn tee-wye fittings, except that short-turn tee-wye fittings may be used on vertical stacks. Fittings other than these may be used if such fittings are approved in accordance with the rules of the authorities. All $\frac{1}{4}$ bends shall be of the long-turn type. Tees and crosses may be used in vent pipes.

What grade or slope is required in horizontal drainage piping?

Answer: Horizontal drainage piping shall be run in practical alignment and at a uniform grade of at least $\frac{1}{8}$ in. per foot for 4-in. pipe and larger, and $\frac{1}{4}$ in. per foot for 3-in. pipe and smaller.

May old house drains and sewers be connected to a new structure?

Answer: Old house drains and sewers may be used for connections to new structures or new plumbing only when such drains and sewers are found, on examination, to conform in all respects to the requirements of the authorities.

What is meant by *fixture unit*, and how was this term derived?

Answer: The unit system was formulated from tests conducted a number of years ago by the subcommittee on plumbing of the Building Code Committee under the Department of Commerce. Standard plumbing fixtures were installed and individually tested, and the amount of liquid waste which could be discharged through their outlet orifices in a given interval was carefully measured.

During the test it was found that a wash basin, which is one of the smaller plumbing fixtures, would discharge waste in the amount of approximately $7\frac{1}{2}$ gals. of water per minute. Since 1 cu. ft. contains 7.4805 gals., it will be observed that this volume was so close to a cubic foot of water that the committee decided to establish it as a basis of the unit system and termed the discharge of the wash basin as one fixture unit. Therefore, one fixture unit represents approximately $7\frac{1}{2}$ gals. of water.

What are the values in fixture units for common plumbing fixtures?

Answer: Table 10-2 is based on the rate of discharge from a wash basin or lavatory as the unit employed to determine fixture equivalents.

What determines the size of waste outlets in fixtures?

Answer: The size of waste outlets in fixtures is determined by the type and number of fixtures installed. Table 10-3 gives the approximate size of waste outlets for various numbers of fixture units.

Table 10-2.   Fixture Unit Values

Fixture	Units
Lavatory or wash basin	1
Kitchen sink	$1\frac{1}{2}$
Bathtub	2
Laundry tub	2
Combination fixture	3
Urinal	3
Shower bath	3
Floor drain	2
Slop sink	4
Water closet	6
One bathroom group (consisting of water closet, lavatory, bathtub, and overhead shower, or water closet, lavatory, and shower compartment)	8
180 square feet of roof drained	1

Table 10-3.   Waste Sizes for Various Fixture Units

Size of waste outlet in fixtures	Number of units
$\frac{1}{2}$ inch, $\frac{3}{4}$ inch, less than 1 inch	$\frac{1}{2}$
1 inch	1
$1\frac{1}{4}$ inches	2
$1\frac{1}{2}$ inches	3
2 inches	$5\frac{1}{2}$
$2\frac{1}{2}$ inches	8
3 inches	15
4 inches	30
5 inches	50
6 inches	80
8 inches	160

What are the requirements for roof extensions of soil and waste stacks?

Answer: Roof extensions of soil and waste stacks or roof vents shall be run at full size at least 1 ft. above any roof pitched at an angle of 30° or more from the horizontal. They shall be full size, at least 5 ft. where the roof is used for any purpose other than weather protection. If the roof terminal of any vent, soil, or waste pipe is within 10 ft. of any door, window, scuttle, or airshaft, the terminal shall extend at least 3 ft. above the opening.

When soil, waste, or vent pipes are extended through the roof, they should be at least 3 in. Pipes smaller than 3 in. should be provided with a proper increaser located just below the roof line.

What are the minimum sizes of individual soil and waste branches?

Answer: Minimum sizes of soil or waste branches to individual fixtures shall be in accordance with Table 10-4. The size of any stack, building drain, or house sewer should be at least that of the largest branch connected to it.

What determines the size of branch soils and wastes in a plumbing system?

Answer: The required size of branch soils and wastes receiving the discharge of two or more fixtures should be determined on the

Table 10-4.   Soil and Waste Branch Sizes for Various Fixtures

Fixture	Branch Size*
Water closet	3 inches
Floor drains	3 inches
Urinal	2 inches
Slop sink	3 inches
Sink, except slop sink	2 inches
Bathtub	1½ inches
Laundry tray	1½ inches
Shower bath	2 inches
Lavatory	1½ inches
Drinking fountain	1½ inches
Dental cuspidor	1½ inches
Sterilizers with ½-inch waste outlet	1½ inches
Combination fixture, laundry tubs, and kitchen sinks .	2 inches

*Branch sizes may vary according to Code adopted in area of use.

Table 10-5. Size of Piping for Branch and Soil Wastes

Maximum number of fixture units permitted	Maximum number of water closets permitted	Diameter of branch (inches)
2	.....	1½
9	.....	2
20	.....	2½
35	1	3
100	11	4
250	28	5

basis of the total number of fixture units drained by branch soils and wastes, in accordance with Table 10-5.

What are the minimum size waste stacks required for water closets?

Answer: No water closet shall discharge into a stack less than 3 in. in diameter.

What are the rules for installation of oil separators?

Answer: When the liquid wastes from any structure consist wholly or in part of volatile, inflammable oil, and an oil separator is required by law, the fixtures receiving such wastes shall be connected to an independent drainage system discharging into such a separator. Every oil separator should have an individual 3-in. vent extending from the top of such separator to the outer air at a point at least 12 ft. above street level. The discharge from the oil separator shall be either independently connected to the sewer or to the sewer side of the building trap.

A fresh-air inlet shall be provided from the drain at the inlet side of the separator to the outer air, and such inlet shall terminate with the open end at least 6 in. above grade. The diameter of the inlet pipe shall be equal to the diameter of such drain, but in any case, such diameter shall be 3 in. or more. The horizontal drain and one riser shall be at least 3 in. in diameter. Risers shall be carried full size through the roof. Oil separators shall be installed in accordance with rules of the authorities.

Explain the difference between "shall" and "may" when used in a plumbing code.

Answer: Shall is a *mandatory* term; may is a permissive term.

What is a wet vent?

Answer: A wet vent is a vent that also serves as a drain.

What is a yoke vent?

Answer: A yoke vent is a pipe connecting upward from a soil or waste stack to a vent stack for the purpose of preventing pressure changes in the stacks.

May waste from an oil storage plant be connected to a public drain or sewer?

Answer: It is unlawful to connect an oil storage plant with any public drain or sewer, or to permit any liquid product of petroleum to escape into any such drain or sewer.

How should drainage of yard areas and roofs be accomplished?

Answer: Yard areas, courts, and courtyards (if paved), together with all roofs, shall be drained into a storm sewer. See Table 10-6.

How should vent pipes be graded?

Answer: Vent and branch vent pipes should be free from drops or sags, or such pipes shall be graded and connected as to drip back by gravity to a soil or waste pipe. Where vent pipes connect to a horizontal soil or waste pipe, the vent branch shall be taken off above the center line of the pipe, and the vent pipe shall rise vertically or at an angle of 45° to the vertical before offsetting horizontally or connecting to the branch, main waste, or soil vent.

Table 10-6. Size of Piping for Storm Water Only

Diameter of pipe (inches)	Maximum Drained Area in Square Feet		
	A Fall, 1/8 inch per foot	B Fall, 1/4 inch per foot	C Fall, 1/2 inch per foot
2	250	350	500
2 1/2	450	600	900
3	700	1000	1500
4	1500	2100	3000
5	2700	3800	5500
6	4300	6100	9000
8	9600	13,000	19,000
10	16,500	24,000	35,000
12	27,000	40,000	56,000

Table 10-7. Vent Stacks and Branches*

Diameter of Pipe (Inches)	Max. Number of Fixture Units	Max. Length in Feet
1¼	1	45
1½	8	60
2	24	120
2½	48	180
3	84	212
4	256	300
5	600	390
6	1380	510
8	3600	750
10	no limit	no limit
12	no limit	no limit

*Figures may vary due to Code adopted in a particular area.

What is the required size of the vent?

Answer: The required size of the vent shall be determined on the basis of the size of the soil or waste stack, the number of fixture units connected to the vent, and the developed length of the pipe, in accordance with Table 10-7. Vents should be at least $1\frac{1}{2}$ in. in diameter. The diameter of every vent stack shall be at least one-half the diameter of the soil or waste stack served. In determining the developed length of vent pipes, the vent stack and branches shall be considered continuous.

Where main stacks are grouped together at the top of a structure into one pipe which extends through the roof, such combined vent shall be at least equal in area to 75 percent of the sum of the areas of the stacks connecting into such combined vent.

Where should main vents be connected?

Answer: Main vents or vent stacks shall connect at their base to the main soil or waste pipe at least 3 ft. below the lowest vent branch. The size of the connection shall be as prescribed earlier. Stacks shall extend undiminished in size above the roof, or shall be reconnected with the main soil or waste stacks at least 3 ft. above the highest fixture branch. Wherever possible, the base of the vent shall receive the wash of the adjoining soil or waste.

What are the rules for offsets in soil, waste, and vent stacks?

Answer: When cast-iron bell-and-spigot pipe is used, offsets in soil and waste stacks above the highest fixture connection and offsets in vent stacks and connections of these stacks to a soil or waste pipe at the bottom, or to the house drain, shall be made at an angle of at least 45° to the horizontal. Where it is impractical because of structural conditions to provide a 45° angle, authorities may permit a reduction in the angle under such conditions as they may prescribe, and when it constitutes a vent extension of the vertical waste from the two fixtures. It shall be installed with a sanitary cross and not closer than 6 in. to the dip of either trap, each trap to be within 2 ft. from the unit vent. This also applies to a horizontal connection if the common vent is taken off at the point of intersection of the fixture branches.

What are the requirements as to materials in plumbing fixtures?

Answer: Plumbing fixtures shall be made of impervious materials with a smooth surface that is easily kept clean. Water closet bowls and traps shall be made of glazed vitreous earthenware, in one piece, and shall be of such form as to hold a sufficient quantity of water when filled to the trap overflow. To prevent fouling of the surfaces, such bowls and traps shall be provided with integral flushing rims constructed to flush the entire interior of the bowl. Urinals shall be made of glazed earthenware.

What are the rules as to location of water closets?

Answer: Outside location of water closets is prohibited. Water closet accommodations shall be placed inside the structures which they serve, except as provided for temporary privies, or privies to be used where no public sewer is available. Whenever a street sewer connection is available, it is unlawful to replace an inside water closet with an outside water closet.

What types of water closets are prohibited?

Answer: It is unlawful to have pans, plungers, offset washout and washout, or other water closets having unventilated spaces or walls which are not thoroughly washed out at each flushing.

What are the rules on flushing and overflow of water closets?

Answer: Every water closet or urinal shall be flushed from a separate flush tank or through an approved flush valve. It is unlawful to connect water closets or urinals directly to a water supply system, except through approved flush valves located to prevent pollution of the water supply. Overflows of flush tanks may discharge into water closets or urinals, but it is unlawful to connect such overflows with any part of the drainage system.

What flush pipe sizes are required for use on water closets?
Answer: Water-closet flush pipes shall be at least $1\frac{1}{4}$ in. in diameter, and urinal flush pipes shall be at least 1 in. in diameter.

What are the rules with regard to the use of antisiphon devices on fixtures?
Answer: Wherever the supply to a fixture is introduced below the overflow level, the supply shall be provided with an approved vacuum breaker that will prevent the siphoning of water from such fixture into the supply piping.

What is a circuit vent?
Answer: A circuit vent is a branch vent that serves two or more traps and extends from in front of the last fixture connection of a horizontal branch to the vent stack.

What are the rules for determining the number of toilets required in a public building?
Answer: Every office building, school, store, warehouse, manufacturing establishment, or other structure where workmen or workwomen are or will be employed, shall be provided with at least one water closet. Water closets shall be provided for each sex according to Table 10-8. The number of water closets to be provided

Table 10-8. Number of Water Closets Required

Number of Persons	Number of Closets	Ratio*
1–15	1	1 for 15
16–35	2	1 for 17½
36–55	3	1 for 18⅓
56–80	4	1 for 20
81–110	5	1 for 22
111–150	6	1 for 25
151–190	7	1 for 27½

*Ratios shown may vary due to Code adopted in a particular area.

for each sex shall in every case be based upon the maximum number of persons of that sex employed at any one time on the given floor, or in the structure for which such closets are provided.

What are the requirements as to location of water closets?

Answer: Water closets should be readily accessible to the persons who will use them. It is usually unlawful to locate water closets more than one floor above or below the regular working place of the persons using them, except that authorities may determine locations in warehouses, garages, and similar structures of low occupancy. The location requirement is inapplicable when passenger elevators are provided for employees to go to the floors where toilets are located.

What are the rules as to installation of traps in plumbing fixtures?

Answer: Each fixture shall be separately trapped as close to the fixture as possible, except that a battery of two or three laundry trays, one sink, and two laundry trays or two compartment sinks may connect with a single trap when the outlets are 2 in. or less. Traps shall be as near to the fixture as possible, but at least within 2 ft. developed length from the outlet. It is unlawful to discharge the waste from a bathtub or other fixture into the water-closet trap or bend. It is unlawful to double-trap fixtures.

What is the required design for fixture traps?

Answer: Traps shall be self-cleaning and water-sealed, and have a scouring action. Traps for bathtubs, lavatories, sinks, and other similar fixtures shall be integral or of lead, brass, plastic, cast iron, or galvanized malleable iron. Traps shall have a full-size bore and a smooth interior waterway such that a solid ball, $\frac{1}{4}$ in. smaller in diameter than the specified diameter of the trap, will pass freely from the outlet end entirely through the seal trap. The minimum diameter given is for the soil or waste branch, except that in the case of water closets, the required minimum is $2\frac{1}{2}$ in. In cases other than fixtures, the size of the trap shall be the same as the size of the discharge pipe connecting thereto. Some codes permit use of plastic traps.

What is the minimum water seal of a fixture trap?

Answer: Fixture traps shall have a water seal of at least 2 in.

What are the rules as to setting and protection of fixture traps?
Answer: Traps should be set true with respect to their water seals and protected from frost and evaporation.

What is the required construction for back-water valves?
Answer: Back-water valves should have all bearing parts made of corrosion-resisting metal, and the valves shall be constructed to insure a positive mechanical seal and remain closed, except when discharging wastes. Back-water valves shall be the approved type.

What types of traps are prohibited in a plumbing system?
Answer: Full "S" traps and bell traps are prohibited. Traps having covers, hand holes, or clean-outs held in place by lugs or bolts acting as interceptors for grease, or similar substances, may be used if such traps are approved by the board. Allowable fixture traps are shown in Fig. 10-2.

What type of clean-out is required in fixture traps?
Answer: Easily accessible clean-outs shall be provided at the foot of each vertical waste, soil stack, or inside leader, on all hand holes of running traps, on all exposed or accessible fixture traps (except earthenware traps), and at each change of direction of hor-

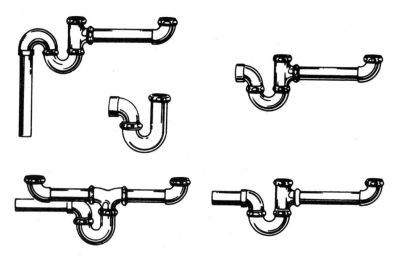

Fig. 10-2. Typical fixture traps.

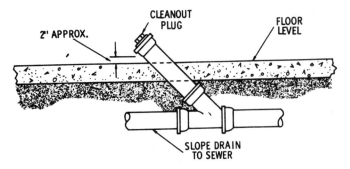

Fig. 10-3.   A typical clean-out in a building drain.

izontal runs. Clean-outs shall be of the same nominal size as the pipes up to 4 in., and such clean-outs shall be at least 4 in. for larger pipes. The maximum distance between the clean-outs in horizontal soil lines shall be 100 ft. A typical sewer clean-out is illustrated in Fig. 10-3.

What is considered a clean-out equivalent in a plumbing system?

Answer: If a fixture trap or a fixture with a trap that is integral can be easily removed without damaging or disturbing the roughing-in work, the device can be designated as a clean-out equivalent— but only if there is not more than a single 90° bend in the line that is subject to be rodded. In a single-story building where sink or lavatory traps are easily removed and are accessible, these traps may be considered as clean-out equivalents.

How should swimming pools be drained?

Answer: Pools should be provided with a drain outlet located so that the entire pool can be emptied. Pools should also be supplied with an overflow at the high-water line, with the drains at least 3 in. in diameter and trapped before connecting with the drainage system. The trap should be vented. Such overflow should be connected to the inlet side of the trap and on the sewer side of the valve on the emptying drain. Drain and circulating outlets should be fitted with a device to reduce the vortex. The spaces around the pool should be drained to prevent the water from draining into the pool. The drains in the gutter may also serve as overflows. The size of the drain and vent connections should be determined by the capacity of the pool when filled to the overflow level. The diameter of the trap should be at least the diameter of the drain pipe.

What method of water circulation should be provided in swimming pools?

Answer: Pools should be equipped to provide a continuous supply of clear, wholesome water at the rate of 20 gals. per hour for each bather using the pool in any one hour. The supply may be either fresh water from an approved water supply system, or such supply may be recirculated if approved means are provided for filtering and sterilizing the water before the water is reintroduced into the pool. The inlets should be located to circulate the water over the entire area of the pool.

The piping of the recirculating system shall be kept entirely separate from the city or domestic supply system. Sterilizing and filtration equipment shall be adequate to keep the pool in a sanitary condition at all times. Adequate shower-bath and toilet accommodations, conveniently located for the use of the bathers, shall be provided for all pools.

Should swimming pools be provided with sterilization and filtration equipment?

Answer: Yes. Sterilizing and filtration equipment shall be adequate to keep the pool in a sanitary condition at all times.

What are the rules as to provisions for shower-bath and toilet facilities in connection with the operation of swimming pools?

Answer: Adequate shower-bath and toilet accommodations, conveniently located for the use of the bathers, shall be provided for all pools.

When does an oil burner pump require a suction line only?

Answer: When the bottom of the oil storage tank is above the burner pump.

When an oil storage tank is installed inside a building, is it necessary to vent it?

Answer: Yes, the vent shall be extended above the roof line of the building.

Why should one end of a buried oil tank be slightly lower than the other end?
Answer: To let any moisture accumulate at the low end.

Should the suction line from a pump be installed at the high or the low end of the tank?
Answer: At the high end.

What size pipe is needed to supply a furnace using 185,000 Btu/hr. if the gas heating value is 1000 and the meter is 40 ft. from the furnace?
Answer: 1-in. pipe.

If a manometer shows a low gas pressure in a building, what is the first item that should be checked?
Answer: The vent from the pressure regulating valve.

At what temperature will propane and butane vaporize?
Answer: Propane vaporizes at $-44°F$, butane at $32°F$.

Why is LP gas more danerous to use than natural gas?
Answer: If LP gas, which is heavier than air, should leak from its container or piping, it will collect in a low place in a room or building or in the ground, needing only a spark to cause a fire or an explosion.

Which gas, propane or butane, is more suitable for winter use in the extreme northern United States?
Answer: Propane.

Where shall LP gas cylinders be stored?
Answer: Outside, in an open building.

A manometer connected to a gas opening shows a pressure of 2.40 on each side of the tube. What is the actual pressure in the piping?
Answer: 4.80 (water column).

What does a symbol such as $\overset{\displaystyle D\,|\,8}{\underline{10}}$ mean on a blueprint?
Answer: It means that a detailed drawing of an area will be found on sheet 10, detail 8.

Why should the plumber and pipe fitter learn to make good isometric drawings?

Answer: A good isometric drawing will show almost every fitting needed on a job.

What does "invert" mean, applied to soil pipe?
Answer: The inside bottom of the pipe.

Checking elevations with an instrument level, with the rod placed on the bench mark, the reading is 6.74. If the bench mark is 820.16, what is the H.I.?
Answer: 826.90

The invert elevation of a pipe is to be at 817.50 and the H.I. of the instrument is 826.90. If the rod is placed on the invert of the pipe, what will be the reading seen through the instrument when the pipe is at the correct elevation?
Answer: 9.40

Why does a plumber need fixture rough-in drawings?
Answer: Rough-in drawings show where the water and waste openings must be "roughed-in," as well as the location of any necessary backing boards or other information in order for the fixtures to be connected properly when the job is finished.

If a job will have approximately thirty 4-in., twelve 3-in., and twelve 2-in. soil pipe joints, how much lead and oakum will be needed?

Answer:

180 lbs. lead	30	12	12	120
	×4	×3	×2	36
18 lbs. oakum	120	36	24	24
				180

Refrigeration and Air-Conditioning Systems

Name the two basic principles that govern all refrigeration systems.

Answer:
1. A liquid absorbs heat when it boils or evaporates to a gas.
2. As vapor or gas condenses to a liquid form, heat is released.

Why is a sight glass installed in the liquid line?
Answer: To indicate a shortage of refrigerant.

When the unit is in continuous operation, what does low pressure on the evaporator side indicate?
Answer: A leaky or stuck expansion valve needle.

If air flow to and from the condenser is blocked in an air-cooled system, what will be the result?
Answer: High head pressure.

Name two causes of a hissing sound at the expansion valve.
Answer: The refrigerant level is too low or the expansion valve needle is stuck open.

Is overcharging a system with refrigerant harmful to the system?
Answer: Yes. Overcharging will cause high head pressure.

What percentage of humidity is generally considered desirable in order to insure comfort?
Answer: From 30 to 50 percent.

What does a warm or hot liquid line indicate?
Answer: A shortage of refrigerant.

Flow is always from a _____ to a _____ pressure area.
Answer: High to low.

What change takes place in the refrigerant when it passes through the evaporator?
Answer: It absorbs heat and is completely vaporized, changing from a liquid to a cold gas.

Why is a compressor necessary in an air-conditioning system?
Answer: The pressure on the refrigerant must be raised so that it can be forced through the condenser and the expansion valve.

What change takes place in the refrigerant as it passes through the expansion valve?

Answer: The refrigerant expands and is partly vaporized; as it expands, it is changed from a hot liquid to a cold liquid plus vapor mixture.

When the refrigerant enters the air-cooled condenser, is the refrigerant hotter or colder than the air entering the condenser?

Answer: Hotter.

Why is a make-up water valve installed in a cooling tower?

Answer: To replace water lost by evaporation.

Steam Heating Systems

Does steam travel faster or slower than other fluid mediums through a heating system?

Answer: Faster.

How can the heat output from steam heating units be raised?

Answer: By raising the steam pressure.

What are the principal advantages of a one-pipe steam system?

Answer: It is dependable and the initial cost is low.

In what type system does the steam and condensate flow in the same direction?

Answer: In a parallel flow system.

An _____ pattern type radiator valve should always be used on a one-pipe steam system. Why?

Answer: Angle, because the steam must enter the radiator, and the condensate must leave through the same valve.

Radiator valves can be used as throttling valves on a one-pipe steam system. True or False? Why?

Answer: False, for two reasons: (1) noise would be created; (2) condensate could not return to the boiler.

What is a Hartford loop and why is it used?

Answer: It is a pressure balancing loop which introduces full boiler pressure on the return side of the boiler to prevent reversed circulation, or water leaving the boiler through the return piping.

In order to be effective, what size should a Hartford loop be? How should it be connected?

Answer:
1. The loop should be the full size of the return main.
2. The horizontal *close* nipple should be installed 2 in. below the boiler water line.

Before steam can enter the piping of a one-pipe system ____ must be eliminated. Why?

Answer: Air, because air in the radiators or piping will block the flow of steam.

Why should an end-of-main vent be installed on large one-pipe steam systems?

Answer: To assure quick venting of the air in a horizontal main.

What is the basic purpose of an air vent?

Answer: It permits the passage of air while blocking the flow of steam and water.

Why is it incorrect to install an end-of-main vent on the last fitting at the end of the steam main?

Answer: Because the high pressure caused by water surge could damage the float in the vent.

What can be done to insure that steam will enter all the radiators in a system at the same time?

Answer: Adjustable-port air vents should be installed on each radiator.

What is the function of a vacuum-type air vent?

Answer: It prevents the return of air, through the vent valve, into the system.

The common definition of a small heating system is one in which the total heat loss is not more than _____.

Answer: 100,000 Btu/hr.

In common practice, the end of the steam supply main should not be less than _____ above the boiler water line.

Answer: 18 in.

On larger systems it is common practice to keep the end of the steam supply main _____ in. above the boiler water line.

Answer: 28.

In a one-pipe steam system to insure the proper flow of steam, air, and condensate, the steam supply and dry return mains should slope (pitch) _____ in the direction of flow of the condensate.

Answer: 1 in. in 20 ft.

How much slope does the wet return require?
Answer: None.

In what type of system does the steam flow in the opposite direction to the flow of condensate?

Answer: In a counter-flow system.

What is the correct pitch for an upfeed riser that is not dripped into the wet return?

Answer: $\frac{1}{2}$ in. per foot.

A downfeed runout should be taken from the (top) (bottom) of the main?

Answer: Bottom.

A one-pipe system with the distribution main installed above the radiators is called a _____ system.

Answer: Downfeed.

What is the function of a float or thermostatic trap?

Answer: It discharges air and condensate into the return while blocking the flow of steam into the return.

Why is a check valve not a workable alternative to a Hartford loop connection?

Answer: Because a foreign object could become lodged under the check valve, preventing it from seating properly.

What does "flash" mean when applied to hot water?

Answer: Water at high temperature, above 210°, will change into steam when the pressure of the water is reduced.

What type of valve should be used for a radiator supply valve in a two-pipe steam system?

Answer: A globe valve.

Why is a cooling leg sometimes necessary when a thermostatic trap is used?

Answer: To permit the condensate to cool sufficiently to open the trap.

Are cooling legs needed for a float and thermostatic trap?
Answer: No.

Steam Heating System Design

What is a vapor system?

Answer: A system that operates at pressures ranging from low pressure to vacuum.

What is the pressure range of a low-pressure system?
Answer: 0 to 15 PSIG.

What is the pressure range of a high-pressure system?
Answer: Pressures above 15 PSIG.

A heating system should be designed to operate at approximately _____ the design load during an average winter.
Answer: One-half.

How should the header on a boiler be sized?

Answer: It should be sized to carry the maximum load that must be carried by any one part of it.

A supply main for a one-pipe system should be at least _____ in size.
Answer: 2 in.

At the point where a supply main is decreased in size, an _____ is the correct fitting to use.
Answer: Eccentric reducer.

What is the advantage in using a two-circuit main with a one-pipe system?
Answer: Quick and uniform delivery of heat.

For a two-pipe system, the minimum pitch or slope for steam and return mains should be not less than _____.
Answer: $\frac{1}{4}$ in. per foot.

What is the minimum pitch for horizontal runouts and risers in a two-pipe system?
Answer: $\frac{1}{2}$ in. per foot.

If the runout is over 8 ft. in length and the minimum pitch cannot be obtained, what must be done?
Answer: The pipe size must be increased to one pipe size larger than called for in the capacity table.

Hot-Water Heating Systems

Name the three ways in which heat is transmitted.
Answer: Conduction, convection, and radiation.

1. Psi means _____?
2. PSIG means _____?
Answer:
1. pounds per square inch.
2. pounds per square inch gauge.

Why does water circulate in a gravity-type system?
Answer: Because hot water is lighter and less dense than cold water, and rises. Cold water, being denser and heavier than hot water, drops, thus establishing circulation.

Why are balancing cocks used in hot-water heating systems?
Answer: To balance the flow, thus insuring that hot water will be forced through each unit of radiation in the system.

What is the principal advantage in a reversed-return system?
Answer: The actual developed length of the supply and return piping to each unit of radiation is the same, resulting in a balanced system.

Why is an expansion tank necessary in a closed hot-water heating system?
Answer: To permit the expansion and compression that takes place as water is heated.

What are the principal advantages in the use of special air control fittings?
Answer: Special air control fittings help eliminate the air from piping and radiation and channel the air into the compression (expansion) tank.

What is the normal workable ratio of water to air in an expansion tank?
Answer: $\frac{1}{3}$ to $\frac{1}{2}$ water; the balance, air.

What is meant by a "waterlogged" expansion tank?
Answer: A tank completely filled with water.

Why are air vents installed in high points of a hot-water heating system?
Answer: On initial start-up of a system, or after draining a system down, air tends to collect or be pushed to the high points in a system. This air must be vented to permit the circulation of water.

What happens in a heating system when an expansion tank becomes waterlogged?

Answer: There is no air for the water to expand or compress against, and the pressure buildup as the water is heated will cause the relief valve to open and discharge the excess water.

Water heated from 40° to 200° will expand approximately _____ percent.

Answer: 4 percent.

A minimum of two aquastats should be installed on a hot-water boiler, one to serve as the _____, one to serve as the _____ control.

Answer: Operating control, high-limit control.

Why should valves be installed on the inlet and discharge sides of circulating pumps?

Answer: To permit removal and replacement or repair of the pump without draining the system.

When radiant heat piping is installed in concrete floors, the maximum water temperature should never exceed _____°.

Answer: 85°.

What is the maximum recommended distance between coils in a radiant heat panel?

Answer: 12 in.

What is the percentage of radiant heat emitted by (a) ceiling panels? (b) floor panels?

Answer: (a) 65 percent. (b) 50 percent.

Which type of coil is best when a constant panel surface temperature is required?

Answer: A grid-type coil.

What is the recommended maximum length of a floor coil using $\frac{1}{2}$-in. tubing?

Answer: 180 ft.

Boiler Fittings

What is the function of a safety (relief) valve on a boiler?
Answer: To open and relieve excess pressure in case of a malfunction.

How does a low water cutoff act to provide protection in an emergency?
Answer: A low water cutoff interrupts the firing circuit of a boiler.

What is the purpose of a stop-check valve?
Answer: When two or more boilers are connected to a common header, a stop-check valve in a boiler header prevents the backflow of steam into the boiler in the event of failure of that boiler.

What is the purpose of a blow-off valve?
Answer: It is used to drain off impurities from the lowest point of the boiler.

Why are water gauges installed on steam boilers?
Answer: To give a visual indication of the water level in the boiler.

Why is it desirable to install a "pigtail" between the steam gauge and the boiler?
Answer: To prevent live steam from damaging the gauge.

Why is an injector used on some steam boilers?
Answer: An injector uses a steam jet to force water into a steam boiler against the boiler pressure.

Why are fusible plugs used on some boilers?
Answer: To protect the boiler in case of a low water condition.

What is the purpose of a steam loop?
Answer: It is used to return condensate to the boiler.

Name two types of pumps used in steam heating systems.
Answer: Condensate pumps and vacuum pumps.

Planning a Heating System

What is the "U" factor?
Answer: A value known as a "coefficient of heat transmission."

What does the U factor represent?
Answer: The time rate of heat flow in Btu/hr. for 1 sq. ft. of surface with a temperature difference of 1°F between the air on one side and the air on the other.

What is the outside design temperature?
Answer: A temperature that will normally be reached during extremely cold weather.

What is the design temperature difference?
Answer: It is the variation between the outside design temperature and the desired indoor temperature.

How is the heat loss determination made?
Answer: List the design temperature difference on the tabulation sheet and multiply the difference by the U factor to determine the heat loss per square foot.

What does EDR mean?
Answer: Equivalent direct radiation.

One sq. ft. of EDR = _____?
Answer: 240 Btu/hr.

If 20,400 Btu/hr. are required to heat a room, how many sq. ft. of radiation are required?
Answer: 85 sq. ft. of EDR capacity.

Where is baseboard radiation normally installed?
Answer: Along an outside wall.

Welding and Brazing

How are brazing temperatures different from welding temperatures?

Answer: Brazing alloy is applied at a temperature below the melting temperature of the metal being brazed. Welding is done at a temperature at or above the melting point of the metal being welded.

What change takes place when malleable iron is heated above 1575°F?
Answer: It reverts to cast iron and is no longer ductile.

Is brazing of stainless steel recommended?
Answer: It is difficult to braze stainless steel with copper-zinc (brass) alloys. Stainless steel can be readily brazed with silver alloys.

Does silver alloy brazing (silver soldering) require the use of flux?
Answer: Yes, in most instances. Copper-to-copper joints do not require flux if the metals are clean and bright.

The end of a pipe should be cut and beveled to an angle of approximately _____° for a welded joint.
Answer: 35°.

How many tack welds should be made when welding a 4-in. joint?
Answer: Four.

In what order should these tack welds be made?
Answer: 1-3; 2-4.

Why should an oxygen cylinder valve be opened fully?
Answer: Because it is a double-seated valve and may leak unless fully opened or fully closed.

Why should an acetylene cylinder valve never be opened more than 1½ turns?
Answer: In order that it may be shut off quickly in an emergency.

Why must oxygen never be brought into contact with grease or oil?

Answer: Because oil or grease may ignite violently if brought into contact with oxygen.

Why is a two-stage regulator the best type to use?

Answer: Because it will deliver constant pressure and require less maintenance.

CHAPTER 11

Safety on the Job

Safety Precautions

Workmen working in areas where there is possible danger of head injuries from impact or from falling or flying objects or from electrical shock and burns should wear protective helmets (hard hats). Plastic helmets, being non-conductors, are preferred over aluminum hard hats.

Ear protective devices should be worn when working in areas of high noise level.

Eye and face protection devices, goggles and safety glasses, should be worn when working in areas where machinery or operations present potential eye or face injury. For example, safety glasses should be used when working with or around grinding machinery to prevent eye injury due to flying particles.

Tools should be in good condition. Tools meant to be sharp should be sharp; impact type tools, chisels, etc., should be kept free of mushroomed heads.

Electric power tools should either be of the approved double insulated type or grounded with approved type wiring devices. Gas and oxygen welding tanks should be securely anchored in an upright position to a post, column, wall, or welding cart. The valves on gas and oxygen tanks used for welding should always be "cracked," opened slightly and then closed, before connecting regulator valves. Before a regulator is removed from the cylinder valve, the valve must be closed and gas or oxygen released from the regulator.

Oxygen cylinders and fittings shall be kept away from oil or grease. Cylinder caps, valves, couplings, regulators, hose, and welding torches and barrels shall be kept free from oil and greasy substances and shall not be handled with oily hands or gloves. Oxygen shall not be directed at oily surfaces, greasy clothes, or employed within a fuel oil or other storage tank or vessel.

Adequate ventilation must be assured when cutting, burning, heating, or welding is being done on materials bearing zinc, lead, cadmium, chromium, mercury, or beryllium.

Ladders with broken or missing steps, broken or split side rails are dangerous and should not be used. Extension ladders should have non-slip devices attached at the foot of the ladder. Working on, under, or near a rickety, poorly set up or unstable scaffold is dangerous. If the scaffolding is not safe, stay away from it. If it is your scaffold, make certain it is in good condition, set up correctly, and on a solid, level base. If you must work near floor and wall openings or open stairwells, be sure that guard rails, handrails, and covers are in place.

Plumbers and pipe fitters are often required to work in ditches or excavations. Cave-ins and sliding banks are hazards that can be eliminated by working in an enclosure such as a four-sided box which can be pulled along to safeguard the working area.

Plumbers and pipe fitters are often required to signal to a crane operator when equipment or material is being set in place. Anyone giving crane signals must know how to signal and must give the correct signal for the specific operation. The signals shown in Figures 11-1 and 11-2 are standard signals used by hoisting engineers and crane operators. Fig. 11-1 shows the signals employed by workers on mobile and locomotive cranes. Fig. 11-2 shows those used on monorails and underhung cranes.

Safety shoes with hardened toe caps should always be worn on construction jobs. There are, of course, many other suggestions and rules for safety on the job. Perhaps the best rule of all is to be safety-conscious, *think* safety. Ninety-nine percent of all job-related injuries are preventable if safety is kept in mind.

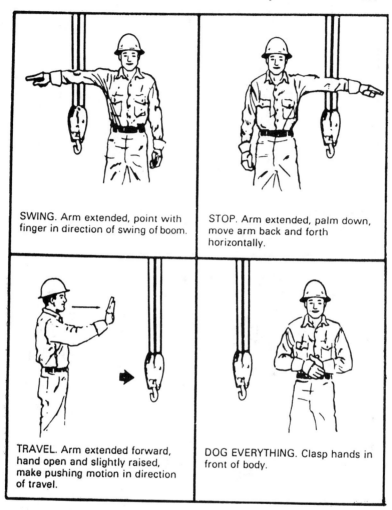

SWING. Arm extended, point with finger in direction of swing of boom.

STOP. Arm extended, palm down, move arm back and forth horizontally.

TRAVEL. Arm extended forward, hand open and slightly raised, make pushing motion in direction of travel.

DOG EVERYTHING. Clasp hands in front of body.

Fig. 11-1. **ANSI/ASME B30.5 Mobile and Locomotive Cranes.** *(Reproduced by permission of the American Society of Automotive Engineers)*

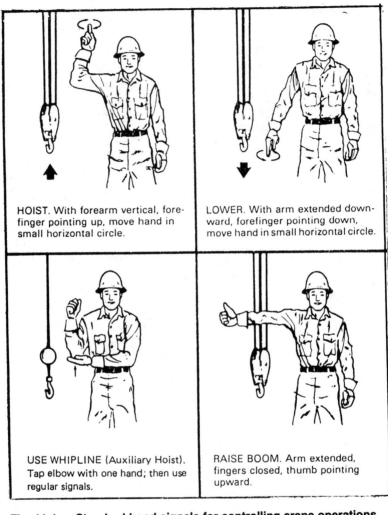

HOIST. With forearm vertical, fore-finger pointing up, move hand in small horizontal circle.

LOWER. With arm extended down-ward, forefinger pointing down, move hand in small horizontal circle.

USE WHIPLINE (Auxiliary Hoist). Tap elbow with one hand; then use regular signals.

RAISE BOOM. Arm extended, fingers closed, thumb pointing upward.

Fig. 11-1. Standard hand signals for controlling crane operations (Cont'd).

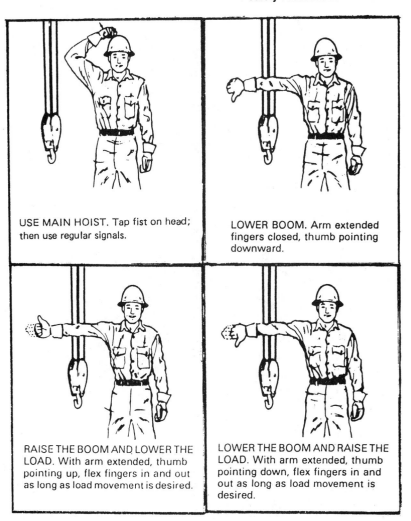

USE MAIN HOIST. Tap fist on head; then use regular signals.

LOWER BOOM. Arm extended fingers closed, thumb pointing downward.

RAISE THE BOOM AND LOWER THE LOAD. With arm extended, thumb pointing up, flex fingers in and out as long as load movement is desired.

LOWER THE BOOM AND RAISE THE LOAD. With arm extended, thumb pointing down, flex fingers in and out as long as load movement is desired.

Fig. 11-1.   Standard hand signals for controlling crane operations (Cont'd).

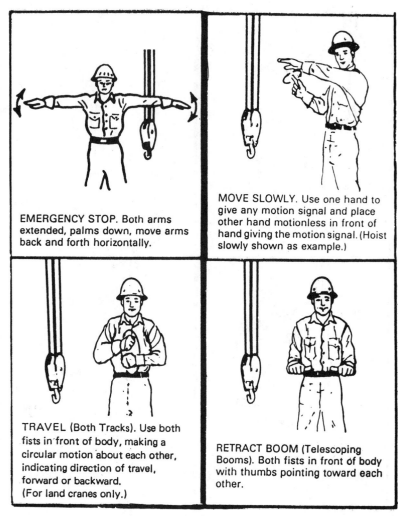

EMERGENCY STOP. Both arms extended, palms down, move arms back and forth horizontally.

MOVE SLOWLY. Use one hand to give any motion signal and place other hand motionless in front of hand giving the motion signal. (Hoist slowly shown as example.)

TRAVEL (Both Tracks). Use both fists in front of body, making a circular motion about each other, indicating direction of travel, forward or backward. (For land cranes only.)

RETRACT BOOM (Telescoping Booms). Both fists in front of body with thumbs pointing toward each other.

Fig. 11-1.   Standard hand signals for controlling crane operations (Cont'd).

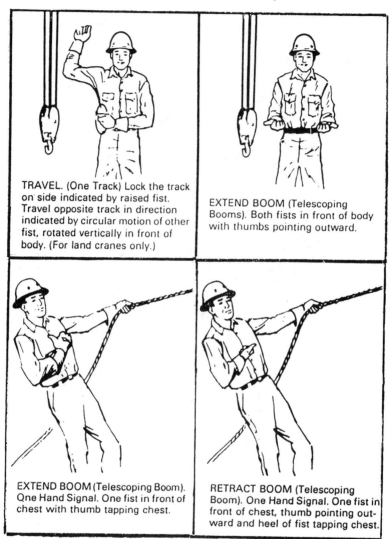

TRAVEL. (One Track) Lock the track on side indicated by raised fist. Travel opposite track in direction indicated by circular motion of other fist, rotated vertically in front of body. (For land cranes only.)

EXTEND BOOM (Telescoping Booms). Both fists in front of body with thumbs pointing outward.

EXTEND BOOM (Telescoping Boom). One Hand Signal. One fist in front of chest with thumb tapping chest.

RETRACT BOOM (Telescoping Boom). One Hand Signal. One fist in front of chest, thumb pointing outward and heel of fist tapping chest.

Fig. 11-1. Standard hand signals for controlling crane operations (Cont'd).

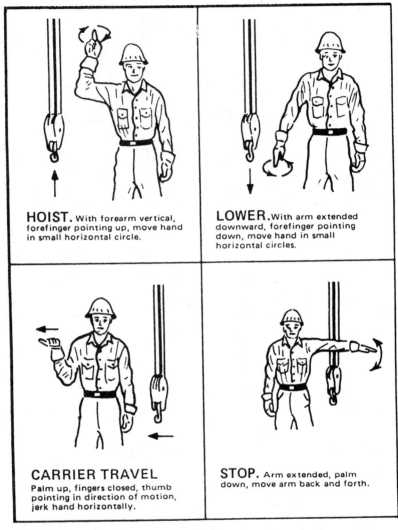

HOIST. With forearm vertical, forefinger pointing up, move hand in small horizontal circle.

LOWER. With arm extended downward, forefinger pointing down, move hand in small horizontal circles.

CARRIER TRAVEL
Palm up, fingers closed, thumb pointing in direction of motion, jerk hand horizontally.

STOP. Arm extended, palm down, move arm back and forth.

Fig. 11-1.   Standard hand signals for controlling crane operations (Cont'd).

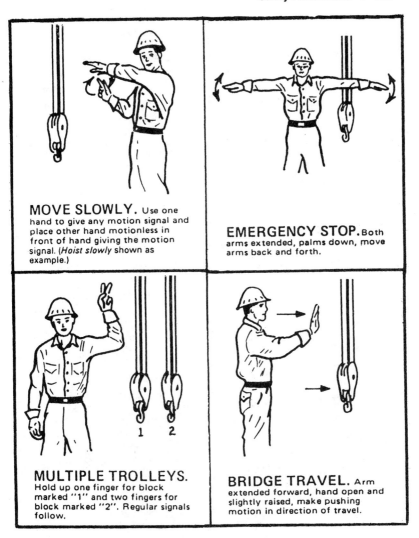

MOVE SLOWLY. Use one hand to give any motion signal and place other hand motionless in front of hand giving the motion signal. (*Hoist slowly* shown as example.)

EMERGENCY STOP. Both arms extended, palms down, move arms back and forth.

MULTIPLE TROLLEYS. Hold up one finger for block marked "1" and two fingers for block marked "2". Regular signals follow.

BRIDGE TRAVEL. Arm extended forward, hand open and slightly raised, make pushing motion in direction of travel.

Fig. 11-2   ANSI/ASME B30.11 Cab-operated Monorails and Underhung Cranes. (*Reproduced by permission of the American Society of Automotive Engineers*)

Other Safety Precautions

Plumbers and pipe fitters are often called on to work in areas where chemicals such as acids, alkalies, or other materials are stored or used. In the event of an accident where acids, caustics, etc., come into contact with any part of the body, especially the eyes, immediate emergency measures are imperative. Chemistry laboratories or other facilities of this nature are required to have installed, in an easily accessible area, safety showers and eyewashes, for use of students or employees. Everyone working in these areas should know exactly where the safety equipment is located and should know how to operate it.

Safety showers and eyewashes are designed with pull chains and handles to enable an injured person, temporarily blinded, to grope for and turn on the water at the shower or eyewash. The eye/face-wash fixture shown in Fig. 11-3 is equipped with a quick-opening, full stay-open flow ball valve, supplying aerated water. Aerated water makes it possible to keep the eyes open longer for more effective rinsing and dilution. The handle is large and easy to locate in an emergency.

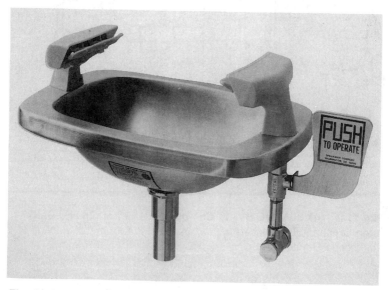

Fig. 11-3. This eye/face-wash unit provides a soft, aerated spray. *(Courtesy Speakman Co.)*

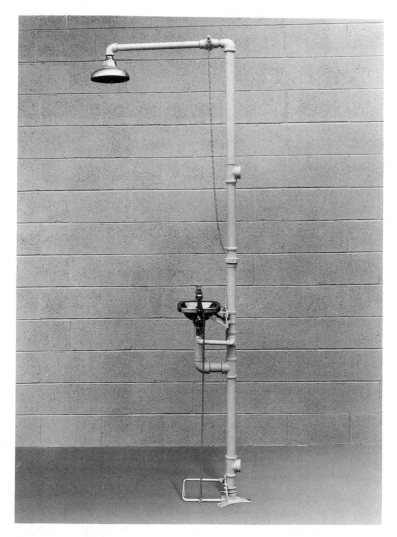

Fig. 11-4.   A combination safety shower and eyewash unit. *(Courtesy Speakman Co.)*

The safety shower, Fig. 11-4, is also equipped with an eyewash. The shower valve is operated by a pull chain. When the valve is opened, an impeller swirls water over the user with a purging action to thoroughly remove hazardous contaminants. The eyewash fixture

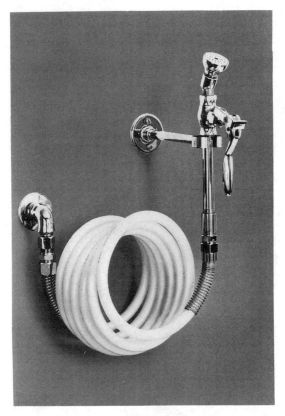

Fig. 11-5. **A hand-held eye/face-wash spray unit.** *(Courtesy Speak-man Co.)*

is operated by a foot pedal, and both units can be operated simultaneously for maximum effect.

The hand-held eye/face-wash unit shown in Fig. 11-5 is more flexible than fixed-station units as it can be directed not only at the eyes but at any part of the body. The valve opens instantly and provides a soft, effective aerated stream of water to the contaminated area.

Potentially hazardous areas such as petroleum fields and construction sites may not have a water source available for the above-mentioned fixtures. A portable unit, shown in Fig. 11-6, contains

Fig. 11-6.   A portable eye-wash/body spray unit. *(Courtesy Speakman Co.)*

an eye, face, body spray, which will supply approximately fifteen minutes of continuous flow of water. This portable unit can provide emergency relief to an accident victim prior to his being transferred to a first aid station or a hospital.

Index

AUDEL®

**Over a Century of Excellence
for the Professional
and
Vocational Trades and the Crafts**

Order now from your local bookstore
or use the convenient order form
at the back of this book.

AUDEL

These fully illustrated, up-to-date guides and manuals mean a better job done for mechanics, engineers, electricians, plumbers, carpenters, and all skilled workers.

CONTENTS

ELECTRICAL

HOUSE WIRING (Sixth Edition)
ROLAND E. PALMQUIST

5 1/2 x 8 1/4   Hardcover   256 pp.   150 Illus.
ISBN: 0-672-23404-1 $14.95

The rules and regulations of the National Electrical Code as they apply to residential wiring fully detailed with examples and illustrations.

PRACTICAL ELECTRICITY
(Fifth Edition)
ROBERT G. MIDDLETON;
revised by L. DONALD MEYERS

5 1/2 x 8 1/4   Hardcover   512 pp.   335 Illus.
ISBN: 0-02-584561-6 $19.95

The fundamentals of electricity for electrical workers, apprentices, and others requiring concise information about electric principles and their practical applications.

GUIDE TO THE 1987 NATIONAL ELECTRICAL CODE
ROLAND E. PALMQUIST

5 1/2 x 8 1/4   Hardcover   664 pp.   225 Illus.
ISBN: 0-02-594560-2 $22.50

The most authoritative guide available to interpreting the National Electrical Code for electricians, contractors, electrical inspectors, and homeowners. Examples and illustrations.

MATHEMATICS FOR ELECTRICIANS AND ELECTRONICS TECHNICIANS
REX MILLER

5 1/2 x 8 1/4   Hardcover   312 pp.   115 Illus.
ISBN: 0-8161-1700-4 $14.95

Mathematical concepts, formulas, and problem-solving techniques utilized on-the-job by electricians and those in electronics and related fields.

FRACTIONAL-HORSEPOWER ELECTRIC MOTORS
REX MILLER and
MARK RICHARD MILLER

5 1/2 x 8 1/4   Hardcover   436 pp.   285 Illus.
ISBN: 0-672-23410-6 $15.95

The installation, operation, maintenance, repair, and replacement of the small-to-moderate-size electric motors that power home appliances and industrial equipment.

ELECTRIC MOTORS (Fourth Edition)
EDWIN P. ANDERSON;
revised by REX MILLER

5 1/2 x 8 1/4   Hardcover   656 pp.   405 Illus.
ISBN: 0-672-23376-2 $14.95

Installation, maintenance, and repair of all types of electric motors.

HOME APPLIANCE SERVICING (Fourth Edition)
EDWIN P. ANDERSON;
revised by REX MILLER

5 1/2 x 8 1/4   Hardcover   640 pp.   345 Illus.
ISBN: 0-672-23379-7 $22.50

The essentials of testing, maintaining, and repairing all types of home appliances.

TELEVISION SERVICE MANUAL (Fifth Edition)

ROBERT G. MIDDLETON;
revised by JOSEPH G. BARRILE

5 1/2 x 8 1/4 Hardcover 512 pp. 395 Illus.
ISBN: 0-672-23395-9 $16.95

A guide to all aspects of television transmission and reception, including the operating principles of black and white and color receivers. Step-by-step maintenance and repair procedures.

ELECTRICAL COURSE FOR APPRENTICES AND JOURNEYMEN (Third Edition)

ROLAND E. PALMQUIST

5 1/2 x 8 1/4 Hardcover 478 pp. 290 Illus.
ISBN: 0-02-594550-5 $19.95

This practical course in electricity for those in formal training programs or learning on their own provides a thorough understanding of operational theory and its applications on the job.

QUESTIONS AND ANSWERS FOR ELECTRICIANS EXAMINATIONS (Ninth Edition)

ROLAND E. PALMQUIST

5 1/2 x 8 1/4 Hardcover 316 pp. 110 Illus.
ISBN: 0-02-594691-9 $18.95

Based on the 1987 National Electrical Code, this book reviews the subjects included in the various electricians examinations—apprentice, journeyman, and master. Question and Answer format.

MACHINE SHOP AND
MECHANICAL TRADES

MACHINISTS LIBRARY
(Fourth Edition, 3 Vols.)

REX MILLER

5 1/2 x 8 1/4 Hardcover 1352 pp. 1120 Illus.
ISBN: 0-672-23380-0 $52.95

An indispensable three-volume reference set for machinists, tool and die makers, machine operators, metal workers, and those with home workshops. The principles and methods of the entire field are covered in an up-to-date text, photographs, diagrams, and tables.

Volume I: Basic Machine Shop
REX MILLER

5 1/2 x 8 1/4 Hardcover 392 pp. 375 Illus.
ISBN: 0-672-23381-9 $17.95

Volume II: Machine Shop
REX MILLER

5 1/2 x 8 1/4 Hardcover 528 pp. 445 Illus.
ISBN: 0-672-23382-7 $19.95

Volume III: Toolmakers Handy Book
REX MILLER

5 1/2 x 8 1/4 Hardcover 432 pp. 300 Illus.
ISBN: 0-672-23383-5 $14.95

MATHEMATICS FOR MECHANICAL TECHNICIANS AND TECHNOLOGISTS

JOHN D. BIES

5 1/2 x 8 1/4 Hardcover 342 pp. 190 Illus.
ISBN: 0-02-510620-1 $17.95

The mathematical concepts, formulas, and problem-solving techniques utilized on the job by engineers, technicians, and other workers in industrial and mechanical technology and related fields.

MILLWRIGHTS AND MECHANICS GUIDE
(Fourth Edition)

CARL A. NELSON

5 1/2 x 8 1/4 Hardcover 1,040 pp. 880 Illus.
ISBN: 0-02-588591-x $29.95

The most comprehensive and authoritative guide available for millwrights, mechanics, maintenance workers, riggers, shop workers, foremen, inspectors, and superintendents on plant installation, operation, and maintenance.

WELDERS GUIDE (Third Edition)

JAMES E. BRUMBAUGH

5 1/2 x 8 1/4 Hardcover 960 pp. 615 Illus.
ISBN: 0-672-23374-6 $23.95

The theory, operation, and maintenance of all welding machines. Covers gas welding equipment, supplies, and process; arc welding equipment, supplies, and process; TIG and MIG welding; and much more.

WELDERS/FITTERS GUIDE

JOHN P. STEWART

8 1/2 x 11 Paperback 160 pp. 195 Illus.
ISBN: 0-672-23325-8 $7.95

Step-by-step instruction for those training to become welders/fitters who have some knowledge of welding and the ability to read blueprints.

SHEET METAL WORK

JOHN D. BIES

5 1/2 x 8 1/4 Hardcover 456 pp. 215 Illus.
ISBN: 0-8161-1706-3 $19.95

An on-the-job guide for workers in the manufacturing and construction industries and for those with home workshops. All facets of sheet metal work detailed and illustrated by drawings, photographs, and tables.

POWER PLANT ENGINEERS
GUIDE (Third Edition)

FRANK D. GRAHAM;
revised by CHARLIE BUFFINGTON

5 1/2 x 8 1/4 Hardcover 960 pp. 530 Illus.
ISBN: 0-672-23329-0 $27.50

This all-inclusive, one-volume guide is perfect for engineers, firemen, water tenders, oilers, operators of steam and diesel-power engines, and those applying for engineer's and firemen's licenses.

MECHANICAL TRADES
POCKET MANUAL
(Second Edition)

CARL A. NELSON

4 x 6 Paperback 364 pp. 255 Illus.
ISBN: 0-672-23378-9 10.95

A handbook for workers in the industrial and mechanical trades on methods, tools, equipment, and procedures. Pocket-sized for easy reference and fully illustrated.

PLUMBING

PLUMBERS AND PIPE
FITTERS LIBRARY
(Fourth Edition, 3 Vols.)

CHARLES N. McCONNELL

5 1/2 x 8 1/4 Hardcover 952 pp. 560 Illus.
ISBN: 0-02-582914-9 $68.45

This comprehensive three-volume set contains the most up-to-date information available for master plumbers, journeymen, apprentices, engineers, and those in the building trades. A detailed text and clear diagrams, photographs, and charts and tables treat all aspects of the plumbing, heating, and air conditioning trades.

Volume I: Materials, Tools, Roughing-In

CHARLES N. McCONNELL;
revised by TOM PHILBIN

5 1/2 x 8 1/4 Hardcover 304 pp. 240 Illus.
ISBN: 0-02-582911-4 $20.95

Volume II: Welding, Heating, Air Conditioning

CHARLES N. McCONNELL;
revised by TOM PHILBIN

5 1/2 x 8 1/4 Hardcover 384 pp. 220 Illus.
ISBN: 0-02-582912-2 $22.95

Volume III: Water Supply, Drainage, Calculations

CHARLES N. McCONNELL;
revised by TOM PHILBIN

5 1/2 x 8 1/4 Hardcover 264 pp. 100 Illus.
ISBN: 0-02-582913-0 $20.95

HOME PLUMBING HANDBOOK
(Third Edition)

CHARLES N. McCONNELL

8 1/2 x 11 Paperback 200 pp. 100 Illus.
ISBN: 0-672-23413-0 $13.95

An up-to-date guide to home plumbing installation and repair.

THE PLUMBERS HANDBOOK
(Seventh Edition)

JOSEPH P. ALMOND, SR.

4 x 6 Paperback 352 pp. 170 Illus.
ISBN: 0-672-23419-x $11.95

A handy sourcebook for plumbers, pipe fitters, and apprentices in both trades. It has a rugged binding suited for use on the job, and fits in the tool box or conveniently in the pocket.

QUESTIONS AND ANSWERS
FOR PLUMBERS
EXAMINATIONS (Second Edition)

JULES ORAVITZ

5 1/2 x 8 1/4 Paperback 256 pp. 145 Illus.
ISBN: 0-8161-1703-9 $9.95

A study guide for those preparing to take a licensing examination for apprentice, journeyman, or master plumber. Question and answer format.

HVAC

AIR CONDITIONING:
HOME AND COMMERCIAL
(Second Edition)

EDWIN P. ANDERSON;
revised by REX MILLER

5 1/2 x 8 1/4 Hardcover 528 pp. 180 Illus.
ISBN: 0-672-23397-5 $15.95

A guide to the construction, installation, operation, maintenance, and repair of home, commercial, and industrial air conditioning systems.

HEATING, VENTILATING, AND AIR CONDITIONING LIBRARY
(Second Edition, 3 Vols.)
JAMES E. BRUMBAUGH

5 1/2 x 8 1/4 Hardcover 1,840 pp. 1,275 Illus.
ISBN: 0-672-23388-6 $53.85

An authoritative three-volume reference library for those who install, operate, maintain, and repair HVAC equipment commercially, industrially, or at home.

Volume I: Heating Fundamentals, Furnaces, Boilers, Boiler Conversions
JAMES E. BRUMBAUGH

5 1/2 x 8 1/4 Hardcover 656 pp. 405 Illus.
ISBN: 0-672-23389-4 $17.95

Volume II: Oil, Gas and Coal Burners, Controls, Ducts, Piping, Valves
JAMES E. BRUMBAUGH

5 1/2 x 8 1/4 Hardcover 592 pp. 455 Illus.
ISBN: 0-672-23390-8 $17.95

Volume III: Radiant Heating, Water Heaters, Ventilation, Air Conditioning, Heat Pumps, Air Cleaners
JAMES E. BRUMBAUGH

5 1/2 x 8 1/4 Hardcover 592 pp. 415 Illus.
ISBN: 0-672-23391-6 $17.95

OIL BURNERS (Fourth Edition)
EDWIN M. FIELD

5 1/2 x 8 1/4 Hardcover 360 pp. 170 Illus.
ISBN: 0-672-23394-0 $15.95

An up-to-date sourcebook on the construction, installation, operation, testing, servicing, and repair of all types of oil burners, both industrial and domestic.

REFRIGERATION: HOME AND COMMERCIAL (Second Edition)
EDWIN P. ANDERSON;
revised by REX MILLER

5 1/2 x 8 1/4 Hardcover 768 pp. 285 Illus.
ISBN: 0-672-23396-7 $19.95

A reference for technicians, plant engineers, and the home owner on the installation, operation, servicing, and repair of everything from single refrigeration units to commercial and industrial systems.

PNEUMATICS AND
HYDRAULICS

HYDRAULICS FOR OFF-THE-ROAD EQUIPMENT (Second Edition)
HARRY L. STEWART;
revised by TOM PHILBIN

5 1/2 x 8 1/4 Hardcover 256 pp. 175 Illus.
ISBN: 0-8161-1701-2 $13.95

This complete reference manual on heavy equipment covers hydraulic pumps, accumulators, and motors; force components; hydraulic control components; filters and filtration, lines and fittings, and fluids; hydrostatic transmissions; maintenance; and troubleshooting.

PNEUMATICS AND HYDRAULICS (Fourth Edition)
HARRY L. STEWART;
revised by TOM STEWART

5 1/2 x 8 1/4 Hardcover 512 pp. 315 Illus.
ISBN: 0-672-23412-2 $19.95

The principles and applications of fluid power. Covers pressure, work, and power; general features of machines; hydraulic and pneumatic symbols; pressure boosters; air compressors and accessories; and much more.

PUMPS (Fourth Edition)
HARRY STEWART;
revised by TOM PHILBIN

5 1/2 x 8 1/4 Hardcover 508 pp. 360 Illus.
ISBN: 0-672-23400-9 $15.95

The principles and day-to-day operation of pumps, pump controls, and hydraulics are thoroughly detailed and illustrated.

CARPENTRY AND
CONSTRUCTION

CARPENTERS AND BUILDERS LIBRARY (Fifth Edition, 4 Vols.)
JOHN E. BALL; revised by TOM PHILBIN

5 1/2 x 8 1/4 Hardcover 1,224 pp. 1,010 Illus.
ISBN: 0-672-23369-x $43.95

Also available as a boxed set at no extra cost:
ISBN: 0-02-506450-9 $43.95

This comprehensive four-volume library has set the professional standard for decades for carpenters, joiners, and woodworkers.

Volume I: Tools, Steel Square, Joinery
JOHN E. BALL; revised by TOM PHILBIN

5 1/2 x 8 1/4 Hardcover 384 pp. 345 Illus.
ISBN: 0-672-23365-7 $10.95

Volume II: Builders Math, Plans, Specifications
JOHN E. BALL; revised by TOM PHILBIN

5 1/2 x 8 1/4 Hardcover 304 pp. 205 Illus.
ISBN: 0-672-23366-5 $10.95

Volume III: Layouts, Foundations, Framing
JOHN E. BALL; revised by TOM PHILBIN

5 1/2 x 8 1/4 Hardcover 272 pp. 215 Illus.
ISBN: 0-672-23367-3 $10.95

Volume IV: Millwork, Power Tools, Painting

JOHN E. BALL; revised by TOM PHILBIN

5 1/2 x 8 1/4   Hardcover   344 pp.   245 Illus.
ISBN: 0-672-23368-1 $10.95

COMPLETE BUILDING CONSTRUCTION (Second Edition)

JOHN PHELPS; revised by TOM PHILBIN

5 1/2 x 8 1/4   Hardcover   744 pp.   645 Illus.
ISBN: 0-672-23377-0 $22.50

Constructing a frame or brick building from the footings to the ridge. Whether the building project is a tool shed, garage, or a complete home, this single fully illustrated volume provides all the necessary information.

COMPLETE ROOFING HANDBOOK

JAMES E. BRUMBAUGH

5 1/2 x 8 1/4   Hardcover   536 pp.   510 Illus.
ISBN: 0-02-517850-4 $29.95

Covers types of roofs; roofing and reroofing; roof and attic insulation and ventilation; skylights and roof openings; dormer construction; roof flashing details; and much more.

COMPLETE SIDING HANDBOOK

JAMES E. BRUMBAUGH

5 1/2 x 8 1/4   Hardcover   512 pp.   450 Illus.
ISBN: 0-02-517880-6 $24.95

This companion volume to the *Complete Roofing Handbook* includes comprehensive step-by-step instructions and accompanying line drawings on every aspect of siding a building.

MASONS AND BUILDERS LIBRARY (Second Edition, 2 Vols.)

LOUIS M. DEZETTEL;
revised by TOM PHILBIN

5 1/2 x 8 1/4   Hardcover   688 pp.   500 Illus.
ISBN: 0-672-23401-7 $27.95

This two-volume set provides practical instruction in bricklaying and masonry. Covers brick; mortar; tools; bonding; corners, openings, and arches; chimneys and fireplaces; structural clay tile and glass block; brick walls; and much more.

Volume I: Concrete, Block, Tile, Terrazzo

LOUIS M. DEZETTEL;
revised by TOM PHILBIN

5 1/2 x 8 1/4   Hardcover   304 pp.   190 Illus.
ISBN: 0-672-23402-5 $13.95

Volume 2: Bricklaying, Plastering, Rock Masonry, Clay Tile

LOUIS M. DEZETTEL;
revised by TOM PHILBIN

5 1/2 x 8 1/4   Hardcover   384 pp.   310 Illus.
ISBN: 0-672-23403-3 $13.95

WOODWORKING

WOOD FURNITURE: FINISHING, REFINISHING, REPAIRING (Second Edition)

JAMES E. BRUMBAUGH

5 1/2 x 8 1/4   Hardcover   352 pp.   185 Illus.
ISBN: 0-672-23409-2 $12.95

A fully illustrated guide to repairing furniture and finishing and refinishing wood surfaces. Covers tools and supplies; types of wood; veneering; inlaying; repairing, restoring, and stripping; wood preparation; and much more.

WOODWORKING AND CABINETMAKING

F. RICHARD BOLLER

5 1/2 x 8 1/4   Hardcover   360 pp.   455 Illus.
ISBN: 0-02-512800-0 $18.95

Essential information on all aspects of working with wood. Step-by-step procedures for woodworking projects are accompanied by detailed drawings and photographs.

MAINTENANCE AND REPAIR

BUILDING MAINTENANCE (Second Edition)

JULES ORAVETZ

5 1/2 x 8 1/4   Hardcover   384 pp.   210 Illus.
ISBN: 0-672-23278-2 $11.95

Professional maintenance procedures used in office, educational, and commercial buildings. Covers painting and decorating; plumbing and pipe fitting; concrete and masonry; and much more.

GARDENING, LANDSCAPING AND GROUNDS MAINTENANCE (Third Edition)

JULES ORAVETZ

5 1/2 x 8 1/4   Hardcover   424 pp.   340 Illus.
ISBN: 0-672-23417-3 $15.95

Maintaining lawns and gardens as well as industrial, municipal, and estate grounds.

HOME MAINTENANCE AND REPAIR: WALLS, CEILINGS AND FLOORS

GARY D. BRANSON

8 1/2 x 11   Paperback   80 pp.   80 Illus.
ISBN: 0-672-23281-2 $6.95

The do-it-yourselfer's guide to interior remodeling with professional results.

PAINTING AND DECORATING

REX MILLER and GLEN E. BAKER

5 1/2 x 8 1/4   Hardcover   464 pp.   325 Illus.
ISBN: 0-672-23405-x $18.95

A practical guide for painters, decorators, and homeowners to the most up-to-date materials and techniques in the field.

TREE CARE (Second Edition)

JOHN M. HALLER

8 1/2 x 11   Paperback   224 pp.   305 Illus.
ISBN: 0-02-062870-6 $9.95

The standard in the field. A comprehensive guide for growers, nursery owners, foresters, landscapers, and homeowners to planting, nurturing and protecting trees.

UPHOLSTERING (Updated)

JAMES E. BRUMBAUGH

5 1/2 x 8 1/4   Hardcover   400 pp.   380 Illus.
ISBN: 0-672-23372-x $15.95

The essentials of upholstering fully explained and illustrated for the professional, the apprentice, and the hobbyist.

AUTOMOTIVE AND ENGINES

DIESEL ENGINE MANUAL
(Fourth Edition)

PERRY O. BLACK;
revised by WILLIAM E. SCAHILL

5 1/2 x 8 1/4   Hardcover   512 pp.   255 Illus.
ISBN: 0-672-23371-1 $15.95

The principles, design, operation, and maintenance of today's diesel engines. All aspects of typical two- and four-cycle engines are thoroughly explained and illustrated by photographs, line drawings, and charts and tables.

GAS ENGINE MANUAL
(Third Edition)

EDWIN P. ANDERSON;
revised by CHARLES G. FACKLAM

5 1/2 x 8 1/4   Hardcover   424 pp.   225 Illus.
ISBN: 0-8161-1707-1 $12.95

How to operate, maintain, and repair gas engines of all types and sizes. All engine parts and step-by-step procedures are illustrated by photographs, diagrams, and troubleshooting charts.

SMALL GASOLINE ENGINES

REX MILLER and MARK RICHARD MILLER

5 1/2 x 8 1/4   Hardcover   640 pp.   525 Illus.
ISBN: 0-672-23414-9 $16.95

Practical information for those who repair, maintain, and overhaul two- and four-cycle engines—including lawn mowers, edgers, grass sweepers, snowblowers, emergency electrical generators, outboard motors, and other equipment with engines of up to ten horsepower.

TRUCK GUIDE LIBRARY (3 Vols.)

JAMES E. BRUMBAUGH

5 1/2 x 8 1/4   2,144 pp.   1,715 Illus.
ISBN: 0-672-23392-4 $45.95

This three-volume set provides the most comprehensive, profusely illustrated collection of information available on truck operation and maintenance.

Volume 1: Engines
JAMES E. BRUMBAUGH

5 1/2 x 8 1/4   Hardcover   416 pp.   290 Illus.
ISBN: 0-672-23356-8 $16.95

Volume 2: Engine Auxiliary Systems
JAMES E. BRUMBAUGH

5 1/2 x 8 1/4   Hardcover   704 pp.   520 Illus.
ISBN: 0-672-23357-6 $16.95

Volume 3: Transmissions, Steering, and Brakes
JAMES E. BRUMBAUGH

5 1/2 x 8 1/4   Hardcover   1,024 pp.   905 Illus.
ISBN: 0-672-23406-8 $16.95

DRAFTING

INDUSTRIAL DRAFTING

JOHN D. BIES

5 1/2 x 8 1/4   Hardcover   544 pp.   Illus.
ISBN: 0-02-510610-4 $24.95

Professional-level introductory guide for practicing drafters, engineers, managers, and technical workers in all industries who use or prepare working drawings.

ANSWERS ON BLUEPRINT READING (Fourth Edition)

ROLAND PALMQUIST;
revised by THOMAS J. MORRISEY

5 1/2 x 8 1/4 Hardcover 320 pp. 275 Illus.
ISBN: 0-8161-1704-7 $12.95

Understanding blueprints of machines and tools, electrical systems, and architecture. Question and answer format.

HOBBIES

COMPLETE COURSE IN STAINED GLASS

PEPE MENDEZ

8 1/2 x 11 Paperback 80 pp. 50 Illus.
ISBN: 0-672-23287-1 $8.95

The tools, materials, and techniques of the art of working with stained glass.